T0140846

Computergestützte Informationsbeschaffung und -verwaltung aus wissenschaftlichen Dokumenten

Computer-aided information retrieval and management system from scientific documents

Zur Erlangung des akademischen Grades eines

DOKTORS DER NATURWISSENSCHAFTEN

(Dr. rer. nat.)

der KIT-Fakultät für Chemie und Biowissenschaften
des Karlsruher Instituts für Technologie (KIT)
genehmigte

DISSERTATION

von

M.Sc. Thanh Cam An Nguyen
aus Hue, Vietnam

Dekan: Prof. Dr. Manfred Wilhelm
Referent: Prof. Dr. Stefan Bräse
Koreferent: Prof. Dr. Ralf H. Reussner
Tag der mündlichen Prüfung: 12.12.2019

Band 86
Beiträge zur organischen Synthese
Hrsg.: Stefan Bräse

Prof. Dr. Stefan Bräse
Institut für Organische Chemie
Karlsruher Institut für Technologie (KIT)
Fritz-Haber-Weg 6
D-76131 Karlsruhe

Bibliographic information published by the Deutsche Nationalbibliothek

The Deutsche Nationalbibliothek lists this publication in the Deutsche Nationalbibliografie; detailed bibliographic data are available in the Internet at http://dnb.d-nb.de

ISBN 978-3-8325-5081-3
ISSN 1862-5681

Logos Verlag Berlin GmbH
Comeniushof, Gubener Str. 47,
10243 Berlin
Tel.: +49 030 42 85 10 90
Fax: +49 030 42 85 10 92
INTERNET: http://www.logos-verlag.de

Für meine Familie,
meine Freunde
und mich.

Probleme kann man niemals mit derselben Denkweise lösen, durch die sie entstanden sind.

- Albert Einstein

Die vorliegende Arbeit wurde in der Zeit vom 15.06.2016 bis 06.11.2019 am Institut für Organische Chemie der Fakultät für Chemie und Biowissenschaften am Karlsruher Institut für Technologie (KIT) Campus Nord unter der Leitung von Prof. Dr. Stefan Bräse angefertigt.

Hiermit versichere ich, die vorliegende Dissertation selbstständig verfasst und ohne unerlaubte Hilfsmittel angefertigt zu haben. Es wurden keine anderen als die angegebenen Quellen und Hilfsmittel benutzt. Die aus Quellen – wörtlich oder inhaltlich – entnommenen Stellen wurden als solche kenntlich gemacht. Die Arbeit wurde bisher weder in gleicher, noch in ähnlicher Form einer anderen Prüfungsbehörde vorgelegt oder veröffentlicht. Ich habe die Regeln zur Sicherung guter wissenschaftlicher Praxis im Karlsruher Institut für Technologie (KIT) beachtet.

Table of Contents

Table of Figures

Summary

The research project focuses on two aspects: the development of *ChemScanner* - a software library that can be used for the extraction of chemical information from scientific documents that contain ChemDraw sketches - and the integration of the *ChemScanner* library into the electronic lab notebook (ELN).

ChemDraw is one of the most well-known chemical drawing software that is used by chemists and researchers in recent years[1]. ChemDraw binary (CDX) or ChemDraw XML-based (CDXML) files are the most common file formats for molecular structure drawings. Their contents can also be found embedded within DOC, DOCX, or XML documents. *ChemScanner* was developed to retrieve graphics and schemes directly from these file formats by extracting and interpreting information created by chemical professionals. The obtained data are processed together with the additional text and values to form chemical reactions and molecules.

The outputs from the *ChemScanner* library can facilitate the reuse of chemical information embedded into various documents used as standard storage and communication instrument in chemical sciences (e.g., theses, publications, or patents). The software aims to support the chemists in their efforts to re-use chemistry research data by providing them missing tools for an automated assembly of reaction data.

ChemScanner processing results can be visualized via the *ChemScanner User Interface*, as the central part of the integration of *ChemScanner* into Open Source ELN Chemotion[2]. Via the *ChemScanner User Interface*, users would be able to manage the uploaded ChemDraw files and their outputs from the *ChemScanner*.

The *ChemScanner User Interface* supports the export to Excel and CML, the direct import of the extracted data to the Chemotion ELN, or the "copy and paste" feature for selected information. Imported data could be searched using substructure searching or similarity search. Computational properties of molecules can also be calculated and visualized within the integration UI. The integration system is an essential process, not only to improve the ease of use and feasibility of *ChemScanner* but also to keep track of the project development process with further development extensions.

Zusammenfassung

Das Forschungsprojekt konzentriert sich auf zwei Aspekte: die Entwicklung von *ChemScanner* - einer Software zur Extraktion chemischer Informationen aus wissenschaftlichen Dokumenten mit integrierten ChemDraw Dateien - und die Integration von *ChemScanner* in das elektronische Laborjournal (ELN).

ChemDraw ist eine der bekanntesten und am weitesten verbreitetsten Programme zum Zeichnen chemischer Strukturen[1]. ChemDraw Binärdateien (CDX) oder ChemDraw XML-basierte (CDXML) Dateien sind die gängigsten Dateiformate für Molekülzeichnungen, deren Inhalt auch in DOC-, DOCX- oder XML-Dokumente eingebettet werden kann. *ChemScanner* wurde entwickelt, um chemische Informationen direkt aus den Grafiken und Schemata direkt dieser Dateiformate zu interpretieren und zu extrahieren. Die gewonnenen Daten werden zusammen mit zusätzlichen Texten und Werten zu Molekülen und Reaktionen verarbeitet.

Die Ergebnisse der *ChemScanner*-Bibliothek können die Wiederverwendung von chemischen Informationen erleichtern, die in verschiedenen Dokumenten eingebettet sind, die als Standard-Speicher- und Kommunikationsinstrument in den Chemiewissenschaften verwendet werden (z.B. Dissertationen, Publikationen oder Patente). Die Software soll Chemiker bei der Wiederverwendung von Forschungsdaten unterstützen, indem sie ihnen fehlende Werkzeuge für eine automatisierte Zusammenstellung von Reaktionsdaten zur Verfügung stellt.

Die Ergebnisse der von *ChemScanner* prozessierten Dateien können über die *ChemScanner* Benutzeroberfläche als zentraler Bestandteil der Integration von *ChemScanner* in das Open Source Chemotion ELN[2] visualisiert werden. Durch die Verwendung der *ChemScanner* Benutzeroberfläche können Benutzer die hochgeladenen ChemDraw-Dateien und deren Ausgaben von ChemScanner verwalten.

Die *ChemScanner-Benutzeroberfläche* unterstützt den Export in Excel und CML Dateiformate, den direkten Import der extrahierten Daten in das Chemotion ELN und für ausgewählte Informationen die "Copy and Paste" Funktion. Importierte Daten können über die Substruktursuche oder die Ähnlichkeitssuche ausfindig gemacht werden. Die rechnerischen Eigenschaften von Molekülen können auch innerhalb der Integrations Benutzeroberfläche berechnet und visualisiert werden. Das Integrationssystem ist ein wesentlicher Prozess, nicht nur um die Benutzerfreundlichkeit von *ChemScanner* zu verbessern, sondern auch den Projektentwicklungsprozess mit weiteren Erweiterungen zu unterstützten.

Chapter 1. Introduction

Chemical databases have been playing an essential role for the researcher in organic chemistry. Chemists use chemical information from knowledge databases, experiences, lab notebooks, and literature for synthesis route planning, drug discovery-development, and prediction of new compounds and properties. In the recent years, with the revolution of artificial intelligence (AI) and machine learning (ML), chemical databases are used as the training data sets for the development of machine learning applications in retrosynthesis and reaction prediction progress. Although there are more and more results and improvements, the most crucial obstacle for every machine learning application is the size of the chemical data sets in organic chemistry.

The chemical information retrieval system has been used to automate the extraction progress of chemical information out of scientific literature, comprises patents, theses, and publications. Legacy printed, non-digital resources can be digitalized by scanning then converted into machine-encoded text using Optical Character Recognition (OCR) techniques. The chemical information retrieval systems then curate the digitalized materials for typical information demands in organic chemistry. In practice, chemical entities (e.g., chemical compounds, chemical families, reactions) and their associated information (e.g., preparation steps, safety information) in documents are recognized.

Krallinger et al. [3] described a chemical information retrieval system as a combination of two major components: textual contents and graphical contents, illustrated in figure 1-1.

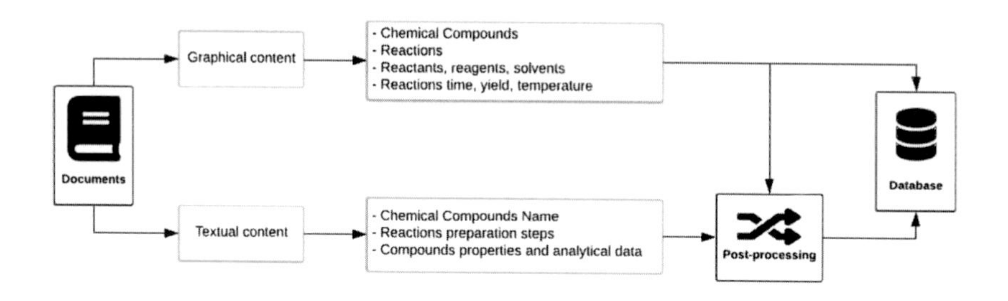

Figure 1-1 The Chemical Information Retrieval System

The *textual contents retrieval component* detects the appearances of chemical entities in the document. Textual chemical entities could be expressed in many methods[4] and do not have a standardized naming convention. Chemical compounds could be described using systematic nomenclature such as IUPAC nomenclature (e.g., dihydrogen monoxide), trivial names (e.g., water), acronyms or abbreviations (e.g., THF, DMF), sum formulas (e.g., C_6H_6), name of groups (e.g., ketones, aldehydes), registered trademarks or brand names (e.g., aspirin, paracetamol). The *textual contents retrieval component* responsibility is to detect and retrieve these entities by employing text mining and natural language processing (NLP) approaches.

Chemical entities that are detected from the *textual contents retrieval component* only have practical meaning if one knows which molecular structures they are referring to. The *post-processing component* is responsible for the linking of chemical entities with the structural information. This can be achieved by employing various approaches. *Martin et al.*[5] employed the search results from chemical databases such as PubChem[6], ChemSpider[7], or SciFinder[8]. *Grego et al.*[9] used a dictionary-based approach for the ChEBI database. The ChemSpot system[10] integrated the name-to-structure software OPSIN[11] to convert IUPAC names to chemical structure. In addition to the name-to-structure conversion, another primary approach for the linking of chemical entities to molecular structures is to use the structural information within documents from the *graphical contents retrieval component*.

Typically, the *graphical contents retrieval component* is an application of OCR[12–14], or specifically in chemistry, is Optical Chemical Structure Recognition (OCSR) to convert image to chemical structure. Although many OCSR applications are developed, reconstructing chemical molecules is an error-prone process[15]. Furthermore, even if the reconstruction is successful, only molecules are rebuilt, the chemical reactions information from images are still missing. Since most chemical professionals are using chemical drawing software (e.g., ChemDraw, ChemSketch, ISIS/Draw …) for their research, the outputs from this software are computer-readable formats so that they would be processed more precisely and reliable.

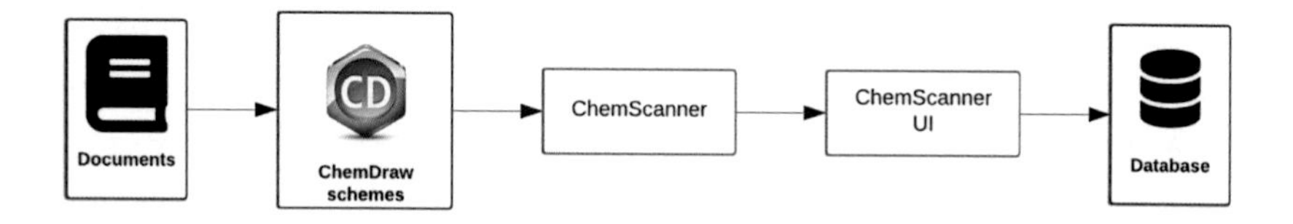

Figure 1-2 Proposed Architecture

To make use of the output from chemical sketcher software, the research project proposes and develops *ChemScanner*, to extract and interpret chemical structural information out from ChemDraw, one of the most prevalent chemical drawing software. The research project also includes the integration of *ChemScanner* into the Open-source Chemotion Electronic Lab Notebook, as shown in figure 1-2. The integration process improves the usability of the software by employing the web application user interface, the *ChemScanner User Interface*. Also, a better management mechanism is provided with the storage management UI together with the searching feature.

1.1 Motivation

The research project is motivated to develop *ChemScanner*, a novel software library that retrieves chemical structural information from output files of the ChemDraw sketcher software.

ChemDraw is known as one of the most popular chemical drawing software. Many chemical researchers and professionals are using ChemDraw for their daily research. However, when their works are shared with others, publicly by publications or internally within their organizations, drawn schemes are treated as digitized images (e.g., BMP, TIFF, PNG, JPG/JPEG or GIF). Its consequences that useful structure-data generated by drawing software is lost, and would not be used properly.

Fortunately, figures drawn by ChemDraw are not entirely disappeared, and there are plenty of resources for mining. Copy and paste figures into Word documents (DOC or DOCX) embed them into the documents while maintaining the original contents of ChemDraw files. Word files (e.g., theses, manuscript) are saved afterward and shared internally or ready to submit for publications. The United States Patent and Trademark Office (USPTO) accepts and stores over 24 million patents ChemDraw files[16].

Current approaches with images to structures techniques (e.g., OSRA[14], CLiDE[17], ChemoCR[18], Imago OCR[19]) are struggling with handle complicated OCR challenges appropriately. For example, OCR software is usually confused and makes many mistakes while dealing with the recognition of wavy bonds, or crossing bonds. Comparing with these approaches, sketches produced by ChemDraw or other drawing software are more computer-readable. By reading the chemical structural contents from the sketches directly, one can overcome many OCR challenges.

For all the above reasons, we believe that our research is a combination of many advantages for the chemical information retrieval system. Our system has the potential to allow more chemical databases to be created. The research project benefits chemical researchers and organizations, as well as the publishers, to mine their data using *ChemScanner* and its integration with the ELN.

1.2 Problem Statement

Although mining sketches seems easy because we can read what users want to draw, without losing any information during translation as OCR approaches, interpreting chemical schemes as what they mean is challenging. Some symbols and graphics are widely used, but their real meaning is entirely different from their original purposes. For example, "*Ar*" purpose is to indicate the chemical element "*Argon*". Nevertheless, in most cases, it is interpreted as "*aryl*". Alternatively, "*Ac*" often means "acetyl" instead of the "*Actinium*" element. More similar ambiguous symbols like "*B*", "*V*", "*W*", "*Y*" intend to represent an ordinary generic atom/group label instead of "*Boron*", "*Vanadium*", "*Tungsten*", "*Yttrium*" accordingly.

Besides, non-chemical elements, including text and graphics, also need to be used in combination with structural information to interpret schemes into proper molecules and reactions. ChemScanner is designed to cover all of these scenarios to correctly derives what they are meant to be. The implementation details are described in chapter 3.

Practically, researchers want to describe the information within the scheme as much as possible. Since there is flexibility for people to use their creativity to create their perfect pattern that fits with their documents, it is almost impossible to reach a perfect conversion without human interaction. In order to solve this problem, the research project also develops the

ChemScanner UI, which is part of the integration of ChemScanner into the Chemotion ELN, to support human monitoring and to improve the conversion time.

1.3 Related Studies

Many commercial solutions introduce sketches mining approaches in the past. Wiley[20] employed the templating approach by defining sets of problems in order to extract molecules with their R-groups label and information. The templates are predefined and are used to extract only molecules. ICSCHEMEPROCESSOR[21] from InfoChem addressed molecular drawing challenges and came up with a hybrid with a templating approach and an algorithmic approach. They are aiming more on molecule extraction than reaction extraction by skipping many drawing patterns of reactions.

Recently, *May et al.*[22] from NextMove Software report a mining approach on the USPTO ChemDraw files, by combining extracted R-group labels and repeated group with R-group table to assemble combinations of molecules from core structures and R-groups. However, they are struggling with undefined chemical entities that cannot be converted.

In summary, these above approaches are only targeting on molecular extraction, without associated reactions information while reactions are more vital to organic chemistry, especially on retrosynthesis and reaction predictions, since reactions are defining they synthetic pathway to desired molecules. Besides, they are all commercial solutions that are closed-source and hard to access by researchers. Also, they do not provide an interface for human interaction with the extraction progress.

1.4 Overview of the research project

In the rest of this thesis, we present our approach to design and implement *ChemScanner* and the integration of *ChemScanner* into Chemotion ELN. The progress is organized as follows.

Chapter 2 will deeply explain the background information about *ChemScanner* development. Basic cheminformatics formats are covered, explaining how molecules and reactions are stored in computer-encoded formats. This chapter also describes how ChemDraw organized the internal sketcher data into CDX and CDXML formats. A review of the current state-of-the-art chemical information retrieval is included in the end.

Chapter 3 presents the overall design and implementation of *ChemScanner*. Challenging issues and particular problems scenario handling are explained in this chapter, together with the use of the library.

Chapter 4 introduces the interface of *ChemScanner*, created as a web application. This chapter cover in detail each feature of the *ChemScanner User Interface* as an intermediate component between *ChemScanner* and the electronic lab notebook.

Chapter 5 describes the integration into the *Chemotion Electronic Lab Notebook* process. This chapter shows every detail of the molecule searching, deployment of the web service used for computational properties, and propose the Green Chemistry attributes.

This thesis is enclosed with chapter 6, which summarizes the development as well as future enhancements.

Chapter 2. Background

This chapter introduces the fundamental concepts that are used for *ChemScanner* and *ChemScanner* development.

First of all, it is explained how molecular information is stored and retrieved with two popular file formats in chemical information (cheminformatics): SMILES and MDL Molfile. Secondly, the details of the ChemDraw file-formats family are described. The formats CDX, CDXML are covered, and it is described how they are embedded inside Word files. Finally, a quick review of current state-of-the-art approaches in chemical information retrieval is given.

2.1 Molecular representing

2.1.1. SMILES

In cheminformatics, a line notation is a single-line string (a sequence of characters), nowadays the most well-known line notation formats are the IUPAC International Chemical Identifier (InChI)[23] and the Simplified Molecular-Input Line-Entry System (SMILES)[24]. InChI is the latest line notation and is more modern than SMILES. However, SMILES is still the best-known because it is more human-readable and is supported by most molecule editors.

The original SMILES specification was introduced by David Weininger[25], then being adapted and modified by many following information systems, especially Daylight Chemical Information System[26]. The latest version of SMILES is an open standard OpenSMILES[27], which was introduced by the Blue Obelisk[28] community.

In general, the SMILES notation is a sequence of characters that end with a whitespace terminator character (space, tab, newline, carriage-return) or the end of the string, while hydrogen atoms could be included or omitted. The SMILES string is obtained by picking one first atom, then printing atomic symbols of atoms while traversing the chemical graph of the molecule structure in any order with five basic rules following.

- **Atoms**: All periodic table elements are supported, asterisk symbol ("*") is accepted as a wildcard or unknown atom.
    - An atom is represented with its respective atomic symbol.

- Upper-case letters indicate non-aromatic atoms, and lower-case letters refer to aromatic atoms. With atomic symbols that have more than one letter, the second letter must be lower-case.
- Atoms do not belong to the organic subset (B, C, N, O, P, S, F, Cl, Br, I), and atoms with abnormal valences must be enclosed in brackets. Figure 2-1 describes typical valences of the organic subset atoms.

- **Bonds**: Single, double, triple, aromatic bonds and disconnected structures are expressed by "-", "=", "#", ":", and "." respectively. Single bonds can be omitted.
- **Branches**: Branched atoms in the chemical graph must be enclosed in parentheses. Branches could be nested or stacked to any depth.
- **Ring**: a number is placed right after the opening and closing ring atoms to identify ring structures with SMILES.
- **Disconnections**: Structures that are disconnected between others are separated in SMILES string by a period symbol (".").

B	C	N	O	P	S	Halogens
3	4	3,5	2	3,5	2,4,6	1

Figure 2-1 Organic subset atoms valences

The same molecule can be represented by different SMILES strings using the basic rules. Figure 2-2 describes four different ways to represent Ethanol. The first form represents Ethanol with implicit hydrogens, and the other three forms represent without implicit hydrogens with a different order of the starting atom. The second and third forms are used more in general since they look simpler and shorter. Therefore, a normalization process is proposed by OpenSMILES, this normalization process is not mandatory, but it is used in many guidelines of chemical information systems to generate SMILES strings.

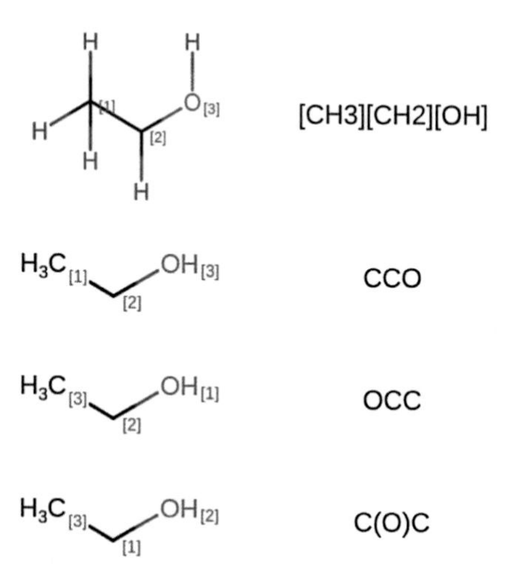

Figure 2-2 Multiple SMILES forms of Ethanol

Standard Form

The standard form of SMILES is the result after applying the normalization process in order to create a compact SMILES. Rules for the normalization process are summarized in figure 2-3. In addition, standardized SMILES string is also considered to be a valid SMILES arbitrary target specification or SMARTS[29], a language for describing molecular patterns. SMARTS allows users to query flexible and efficient substructure-search.

Atoms		
Corrected	Wrong	Rules
CC	[CH3][CH3]	Organic subsets atoms as bare symbols whenever possible
[CH3-]	[CH3-1]	No number if the charge is +1 or -1
C[13CH](C)O	C[13CH1](C)O	No number if hydrogen count equal to 1
[CH3-][C@H](Br)Cl	[C-H3][CH@](Br)Cl	Atom properties order: Chirality, hydrogen-count, charge
[CH3-]	[H][C-]([H])[H]	Group hyrogens with heavy atoms
Bonds		
Corrected	Wrong	Rules
Cc1ccccc1	C-c:1:c:c:c:c:c:1	Skip single and aromatic bonds
Branches		
Corrected	Wrong	Rules
OCc1ccccc1	c1cc(CO)ccc1	Start from terminal atom and heteroatom if possible
CC(C)CCCCCC	CC(CCCCCC)C	Follow the longest branch
Rings		
Corrected	Wrong	Rules
c1ccccc1	C1=CC=CC=C1	Prefer aromatic form than Kekulé form
c1ccccc1C2CCCC2	c0ccccc0C0CCCC0	Start ring number from non-zero, make ring number unique
Disconnections		
Corrected	Wrong	Rules
CC	C1.C1	Only use dots for disconnected structures.

Figure 2-3 Normalization summary

Canonical SMILES

A canonical SMILES is a standard form SMILES that always gives the same SMILES string to any particular molecule, regardless of the source molecule order and position. Benefits from this property, canonical SMILES is employed by many chemical database systems to create unique SMILES for each molecule across the database, store them, and query the input molecule throughout the database precisely.

Canonical SMILES is generated by the canonicalization algorithms, and various canonicalizer have been developed. Notably, among them are the popular open-source cheminformatics software RDKit[30], the chemistry development kit CDK[31], and the chemistry toolbox OpenBabel[32]. OpenBabel uses a canonical algorithm based on InChI[33], CDK canonizing molecules by using extended valence sequences[34], while RDKit uses a standard stable-sorting algorithm[35]. Due to the existence of many canonical algorithms, SMILES and canonical SMILES is considered as not an ideal format for a universal identifier across databases.

2.1.2. Molfile

One of the earliest chemical file formats is the connection table (CTAB) format. CTAB file formats consist of MDL Molfiles, RGfiles, Rxnfiles, SDFiles, RDFiles, XDFiles, and Clipboard[36]. MDL Molfile is a file format that contains molecular spatial information, constituted by information headers, a table enumerating positions of atoms, and a connection table enumerating the bonds that connecting atoms.

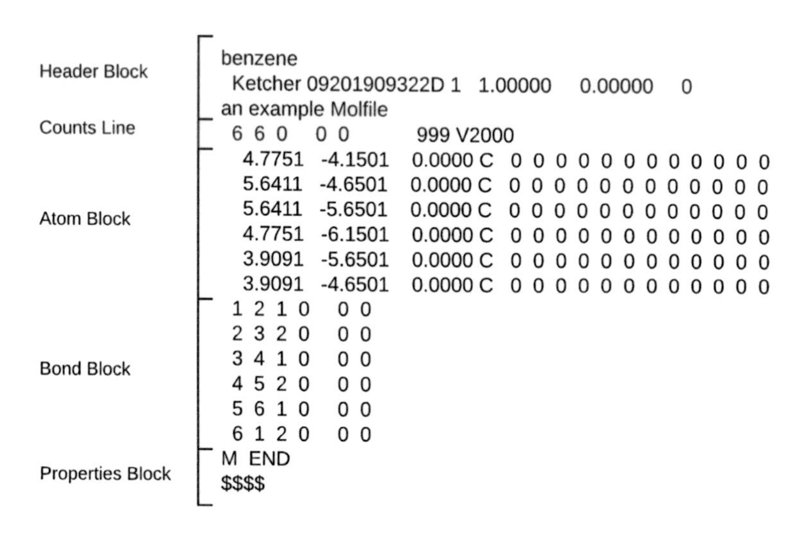

Figure 2-4 Benzene Molfile

There are two versions of MDL Molfile: the V2000 was released earlier, followed by the latest version V3000. A significant limitation of V2000 is that V2000 can only support up to 999 atoms or bonds, while V3000 can support more than 999 atoms or bonds, which extend the usability of Molfile with polymer-based compounds.

Figure 2-4 describes the contents of a benzene's Molfile as an example. The Molfile is created by Ketcher[37], which is employed by Chemotion ELN as an option for chemical structure editor. The Molfile is made of five blocks:

1. **Header**: consist of three lines
    - Line 1 describes the molecule name.
    - Line 2 stores user name, program, created date, and other information.
    - Line 3 is reserved for comments. A blank line is given if there is no comment.
2. **Counts Line**: contains a number of atoms, number of bonds, number of atom lists, chiral flag, and the CTAB version. For example, the "*counts line*" in figure 2-4 shows 6 atoms, 6 bonds, no atom lists, the chiral flag is off, and the version is V2000.
3. **Atom Block**: one atom is described per line, defined by this format: `xxxxx.xxxxyyyyy.yyyyzzzzz.zzzz aaaddcccssshhhbbbvvvHHHrrriiimmmnnneee`
   Characters in the format are explained as follows
    - `x y z`: atom coordinates
    - `aaa`: atom symbol
    - `dd`: mass difference
    - `ccc`: charge
    - `sss`: atom stereo parity
    - `hhh`: explicit hydrogen count + 1
    - `bbb`: stereo care box
    - `vvv`: valence
    - `HHH`: H0 designator
    - `rrr`: reserved (not used)
    - `iii`: reserved (not used)
    - `mmm`: atom-to-atom mapping number
    - `nnn`: invertion/retention flag
    - `eee`: exact change flag

4. **Bond Block**: similar to atom block, one bond is described by one line by following the format: `111222tttsssxxxrrrccc`, which are explained below

    - `111`: the first atom number index
    - `222`: the second atom number index
    - `ttt`: bond type, one of the following values

        - 1: Single bond
        - 2: Double bond
        - 3: Triple bond
        - 4: Aromatic bond
        - 5: Single or Double
        - 6: Single or Aromatic
        - 7: Double or Aromatic
        - 8: Any

    - `sss`: bond stereo, one of the following values

        - 0: not stereo
        - 1: up
        - 6: down
        - 3: cis or trans (either) double bond
        - 4: Either

    - `xxx`: reserved (not used)
    - `rrr`: bond topology: 0 - Either, 1 - Ring, 2- Chain
    - `ccc`: reacting center status

5. **Properties Block**: Most lines in this block follows the format "`M  XXX`". The block ends with "`M  END`".

There are more features of the Molfile that are not covered here, such as structural text descriptor (Stext), the atom list block, advanced syntaxes in properties block (e.g., Rgroup, Sgroup, charge, radical, isotope).

For the scope of this dissertation, a more in-depth explanation of these features is not necessary. In general, the Molfile covers most of the properties of common molecules, and this format can be read and written by most of the cheminformatics software in organic chemistry.

2.1.3. Molecular fingerprints

Structure search has been known as an essential application of informatics in chemistry, digitalization and storing molecule structures have no meaning without the ability to search and retrieve target molecules, with their similarities and substructures or superstructures.

In high-speed structural screening, which is covered in detail in the molecule searching section of chapter 5, molecular fingerprints are employed for the screening process of substructure and similarity search.

Many modern fingerprints are inherited from the fragment-based fingerprint introduced by Daylight Chemical Information System[26], which generate molecular fingerprints by following these steps:

- **Step 1**: generate patterns for each atom. (0-bond paths)
- **Step 2**: generate patterns for each group of atom and others that connect directly to it. (1-bond paths)
- **Step 3**: generate patterns for each group of atoms and bonds by following paths that have 2 bonds. (2-bond paths)
- **Step 4**: Repeat the process with paths that have 3, 4, 5, 6, 7 bonds.
- **Step 5**: apply a pseudo-random number generator, or hash function, to produce a number or a set of numbers.
- **Step 6**: assemble numbers from step 5 to form a fingerprint.

For example, applying the fingerprint generation on molecule `CC(N)=O`, one receives the following patterns:

- 0-bond: `C C N O`
- 1-bond: `CC C=O CN`
- 2-bond: `CC=O CCN O=CN`
- No 3-bond since the longest branch has only 2 bonds.

Similar to canonical SMILES, hash functions would be varied on different systems. The FP2 Fingerprint from OpenBabel is used for molecular searching, that is introduced in chapter 5. The details of the FP2 Fingerprint from OpenBabel is covered in the next section.

OpenBabel's FP2 Fingerprint

FP2 fingerprint is the default fingerprint format in OpenBabel and is used as an option for similarity searching and substructure searching of compounds in Chemotion ELN database.

OpenBabel defined FP2 as a path-based fingerprint, which is similar to Daylight's fingerprint that we explained above, with some modifications:

- Single-atom fragments of O, C, and N are ignored.
- If its atoms form a ring, a fragment would be terminated.
- Patterns (or fragments) would be generated up to 7 atoms.
- In case there are similar fragments, only a single canonical fragment is kept.
- FP2 fingerprint is a 1024-bits vector.

Figure 2-5 describes an example of FP2 fingerprint generating with OpenBabel, on *benzamide* (`NC(=O)C1=CC=CC=C1`). All generated patterns are grouped on the number of bond paths, from 1-bond paths to 6-bond paths: line 1 contains 1-bond patterns, line 2 contains 2-bond patterns, and so on.

The number under each pattern, in figure 2-5, is the output from the FP2 hash function for that pattern. These numbers also indicate the position of the set bit in the 1024-bits space of FP2. *Benzamide* has 25 patterns so that we have 25 bit is set (bit equal to 1) in the fingerprint vector. Those bit positions are specified by the output number of the hash function. For example, 1-bond paths have four numbers: 81, 670, 82, and 623; it means bit at these positions: 81st, 670th, 82nd, and 623th are set to 1.

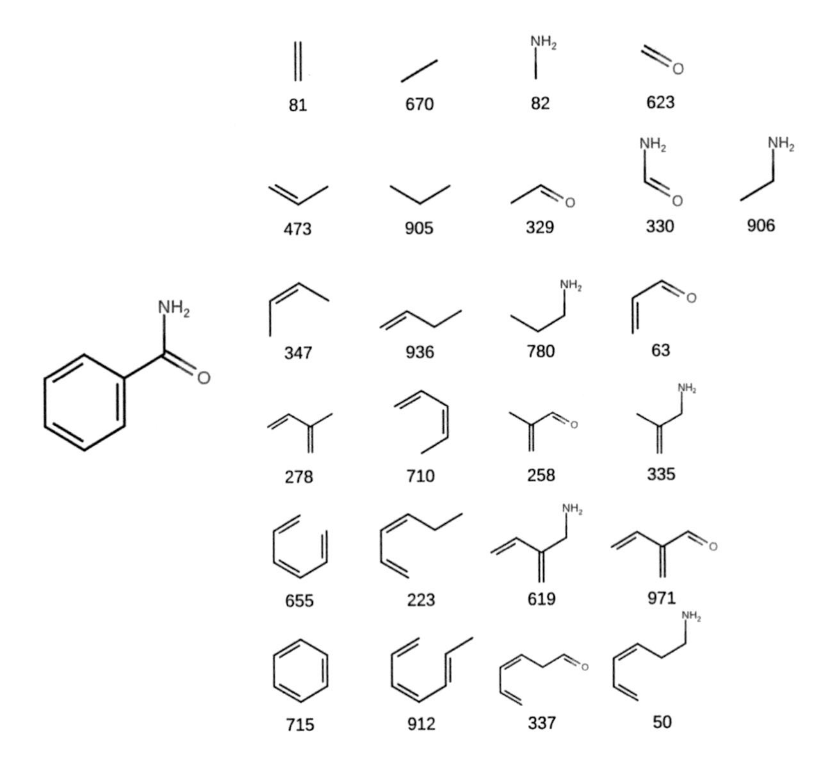

Figure 2-5 Visualization of FP2 fingerprint of Benzamide

Details of the hash function of FP2 can be located in OpenBabel's implementation of FP2 fingerprint[38]. In general, each fragment is represented by the following format

`R A B A` ...

Where **`R`** is zero if fragment is linear, or the order of the bond for cyclic fragments, **`A`** stands for the atom's atomic number, and **`B`** is for the bond order. For example:

- Fragment **`670`** is represented as "**`0 6 1 6`**", which describes a linear fragment of two Carbon atoms that are connected by a single bond. The first zero-digit specifies there is no bond between the last atom and the first atom.

- Fragment **715** is represented as "**5 6 5 6 5 6 5 6 5 6 5 6**". The string describes a ring fragment of six carbon atoms. These carbon atoms are connected by aromatic bonds. The first digit is five indicates that the last atom and the first atom are connected by an aromatic bond.

Each numbers in the representor described above takes part in the hash calculation, a modified of the OpenBabel's hash function[38] is illustrated by figure 2-6

```
unsigned int Hasn(const vector<int> fragment) {
  const int MODINT = 108; // 2^32 % 1021

  unsigned int hash=0;
  for(unsigned i = 0; i < fragment.size(); ++i)
    hash= (hash*MODINT + (fragment[i] % 1021)) % 1021;

  return hash;
}
```

Figure 2-6 OpenBabel FP2 hash function

A pseudo-simulation of fragment 670, known as "**0 6 1 6**", would be:

1. **Initial:** `hash = 0`
2. "**0**": `hash = (0 * 108 + (0 % 1021)) % 1021 = 0`
3. "**6**": `hash = (0 * 108 + (6 % 1021)) % 1021 = 6`
4. "**1**": `hash = (6 * 108 + (1 % 1021)) % 1021 = 649`
5. "**6**": `hash = (649 * 108 + (6 % 1021)) % 1021 = 670`

After the hash calculation, we receive the expected number "**670**" as shown in figure 2-5. The same procedure is applied for all 25 fragments. Afterward, a list of 25 positions is obtained. Each value in this list indicates which bit position is set to 1.

2.2 ChemDraw related formats

As described in chapter 1, our research project is aiming at sketches mining, especially sketches that were drawn by the ChemDraw software. Therefore, understanding the ChemDraw file formats, about how ChemDraw stored the molecular structure into the documents, is an essential background for any further development.

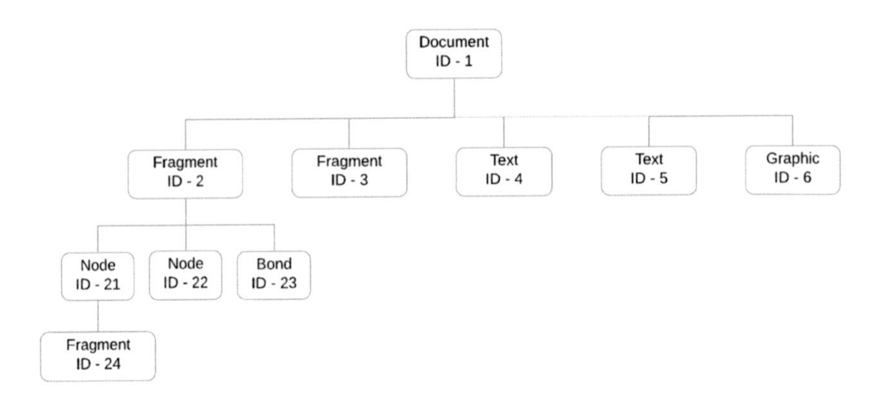

Figure 2-7 Basic structure of ChemDraw formats

The ChemDraw native format, ChemDraw binary format (CDX), is a tagged file format, the CDX file format contains sets of objects with each object is provided with a tag to identify what it represents (e.g., molecular fragment, atom, bond). CDX objects are tagged by hex bytes, while CDXML objects are tagged by strings. Tagged file formats are flexible, tags that are not interested in programs could be effectively ignored during processing, and new tags could be added for expansion without corrupting current existing software systems.

In ChemDraw formats, there are three kinds of tagged items: *object headers*, *properties*, and *object ends*. Every *object header* must have its corresponding ending tag. Otherwise, the file structure would be broken. An object is defined by its header and ending tag, the content within are *properties* of the object or other *objects*.

Figure 2-7 provides a general overview of the basic structures of ChemDraw formats. Stand at the root is the "`Document`" object, the top-level object of ChemDraw formats, that contains all other objects and attributes. In this example, "`Document`" is identified as a tagged item which has an ID equal to 1, and contains two "`Text`" objects, one "`Graphic`" object, two "`Fragment`" objects. `Fragments` in ChemDraw formats are considered as molecular fragments, containing atoms (as "`Node`" object) and bonds (as "`Bond`" object), as shown in figure 2-7.

Many details such as "`Fragment`" properties, "`Atom`" and "`Bond`" properties are omitted in figure 2-7. In the next part, more specific details in CDX and CDXML formats are explained.

2.2.1. ChemDraw Binary Format (CDX)

A typical CDX file is consisted of:

1. A general 28-bytes header.
    - 8 fixed bytes: `"VjCD0100"(56 6A 43 44 30 31 30 30)`
    - 4 reserved bytes `(04 03 02 01)`
    - 16 reserved bytes, all zeros.
2. Objects and properties that are stored as little-endian byte order.
3. The file ending, represented by two bytes of zero `(00 00)`.

In CDX format, objects are composed of

- **Tag identifier**: a two-bytes (16 bits) value, the first most significant bit is always set to 1 in order to differentiate with the tag identifier of properties. The second most significant bit is zero for ChemDraw predefined objects, and equal to 1 for user-defined objects. Therefore, the number of ChemDraw predefined objects we could have is $(2^{14} - 1) = 16383$, and $2^{14} = 16384$ objects for user-defined.
- **Object identifier (ID)**: a four-bytes value immediately follows the tag. Practically, the identifier should be unique within the document scope,
- **Object ending**: the two-bytes value of zero.
- **Object contents**: the contents from object identifier till object end.

Similarly, CDX properties are comprised of

- **Tag identifier**: a two-bytes (16 bits) value, the first most significant bit is always clear (equal to 0). The setting of the second most significant bit is the same as the object's tag identifier.

- **Length**: a two-bytes value indicates the length of the property's data.
- **Data**: a varied set of bytes whose length is specified by the property's length field.

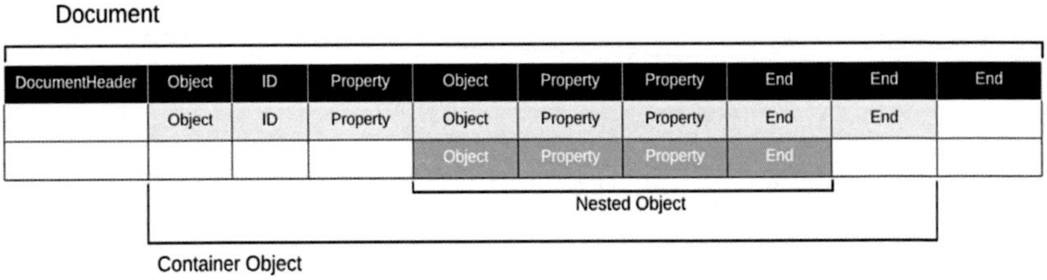

Figure 2-8 An example of a simple document

Considering a simple *document object* that is shown in figure 2-8, the figure describes how the *document object* is structured in the CDX format at the abstract level. In this example, "*Object*" represents the object's tag identifier. Although the raw binary is shown in the first line, nested objects are grouped and described in a separated line for better understanding.

The *document* object in figure 2-8 contains one "*container object*", as represented in the second line, followed by its object identifier (ID) and one property. The "*container object*" then again contains another "*nested object*", with its ID and property, which is illustrated in the third line. The object nesting could be enclosed at any depth level, and typical CDX files are comprised of multiple nested layers and objects, with no limitations on how many objects or properties can be nested on any level.

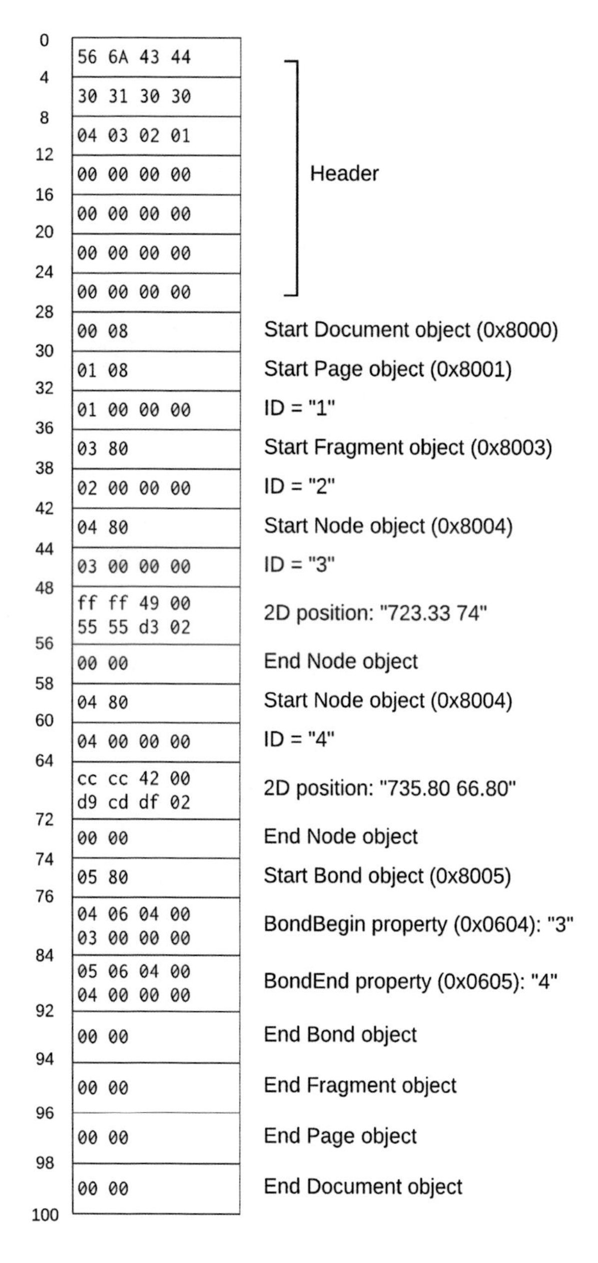

Figure 2-9 A simplified CDX containing an ethane (C2H6) molecule

Figure 2-9 illustrates a CDX file containing an ethane compound (C_2H_6), at the raw binary detail level, the file has a size of 100 bytes. The information within binary is simplified. Objects and properties such as font table, color table, bounding box are ignored in the figure. The above CDX file is broken down as follows

- One `Document` object, identified via the binary tag "`0x8000`". "`Document`" object must contain at least one "`Page`" object.
- One `Page` object, as a child of the `Document` object above. "`Page`" object is represented in binary as "`0x8001`". The object has an identifier equal to `1`.
- The `Page` object mentioned above contains one `Fragment` object, is known via its tag "`0x8003`", which has an ID equal to *2*. "`Fragment`" objects are interpreted as molecular fragments.
- The nested `Fragment` has 3 nested objects:
    - Two `Node` objects, interpreted as molecular atoms, are nested inside the `Fragment` object, which their IDs are 3 and 4. The "`Node`" object is identified with the binary "`0x8004`". "`Node`" object usually contains one property "`Node_Element`" via tag "`0x0402`" to indicate the atomic number of the atom. In case this property is absent, the "`Node`" is considered as a carbon atom.
    - One `Bond` object, represented by the tag "`0x8005`", which does not have an identifier since ID is not mandatory. The `Bond` object connects two `Atom` objects above by using the "`BondBegin`" property to indicates atom at the beginning of the bond, the "`BondEnd`" property pointing to the atom at the bond ends, and the "`Bond_Order`" property to specify the order of the bond. "`BondBegin`" and "`BondEnd`" properties values refer to objects that have the identifier equal to their values. In this example, two atoms above are connected to each other with the order of 1, since "`Bond_Order`" is missing. Therefore, the bond order is set to the default value 1.

2.2.2. Embedded ChemDraw in Microsoft Word documents

One of the most crucial features of ChemDraw is the ability to embed the output sketches into Microsoft Word documents, including Word binary file format (DOC)[39] and Word Extensions to the Office Open XML file format (DOCX)[40]. In this part, an overview at how ChemDraw information is embedded within Word documents is given.

ChemDraw in Word binary file format (DOC)

The Word binary file format (DOC) is used by Microsoft Word 97, Microsoft Word 2000, Microsoft Word 2002, and Microsoft Office Word 2003. A Word Binary File is a Compound File Binary file format (CFB)[41], which can store multiple files and streams within a single file, by partitioning the file into sectors. Stored data in CFB file formats are structured as *storage* and *streams*.

In general, a CFB file can be considered as a container of storages and streams. For the purpose of this thesis, we only cover *storages* and *streams* that are used to extract the embedded CDX data as following

- **`WordDocument`** Stream: The *`WordDocument`* stream is the mainstream and a mandatory stream, that must be present in the file. The mainstream is consisting of two parts: header part and text part. The header, called *`File Information Block`* (`FIB`), contains information of the file and pointers to text part or other streams. The text part holds information of all the text in the document, but not in the order as it is displayed. Meaning the text here is fragmented.
- **`Table Stream`** (*`0Table`* or *`1Table`*): sometimes called *piece table*. *`Table stream`* positions of characters in the file. Text displayed in Microsoft Word application is the combination of the information of the *piece table* with both the header part and text part from the *`WordDocument`* stream.
- Information Summary Stream (*`SummaryInformation`* Stream and *`DocumentSummaryInformation`* Stream): The summary information of the document is saved here, such as title, author, created date, last saved by, number of pages, number of words, number of characters.

- **`Data Stream`**: store all non-text information data (e.g., pictures, embedded data). Data contents in this stream can be accessed via metadata saved in other streams and storages.
- **`ObjectPool Storage`**: contains OLE (Object Linking and Embedding) objects.
    - `ObjInfo` Stream: each storage in `ObjectPool` storage contains an `ObjInfo` stream, holding information about the embedded OLE object.

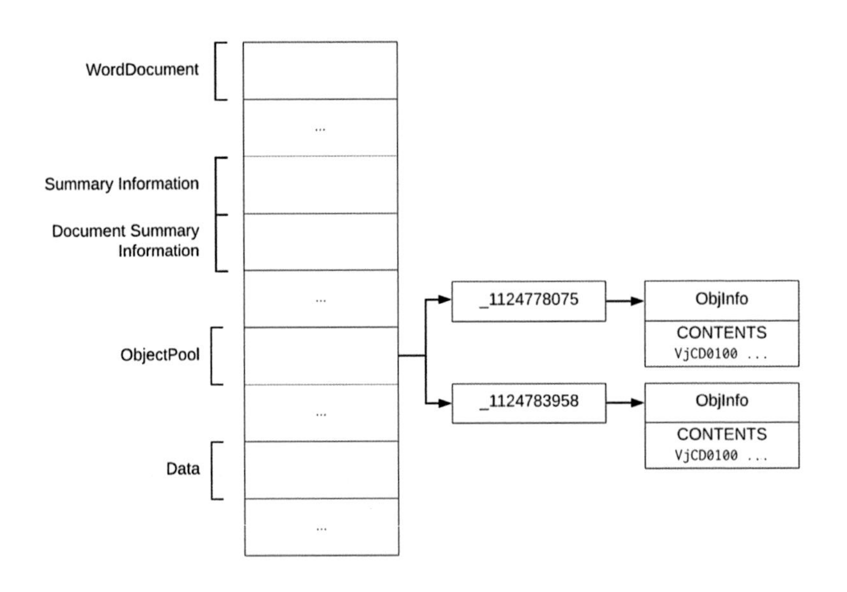

Figure 2-10 Retrieving ChemDraw contents from Word Binary format

The figure above illustrates how a Word file is structured in binary format, and how to retrieve the embedded CDX data, which are saved as OLE objects within the DOC file. In order to retrieve embedded data contents, one needs to get the references to the contents by parsing and reading `ObjectPool` storages. In the example of figure 2-10, the `ObjectPool` storage contains another two storages. Each of them has one `ObjInfo` stream, which refers to the content position and the length of content in the `Data` stream. The embedded ChemDraw would be appropriately extracted by using these positions and contents length.

ChemDraw in Word DOCX file format

DOCX is an adaption of Microsoft Word to the Office Open XML (OOXML) file format[42]. According to the OOXML format, DOCX file format is a zipped archive, XML-based file format. Since XML, or Extensible Markup Language, is designed for both computer and human-readable[43], the DOCX file format is more straightforward than DOC format.

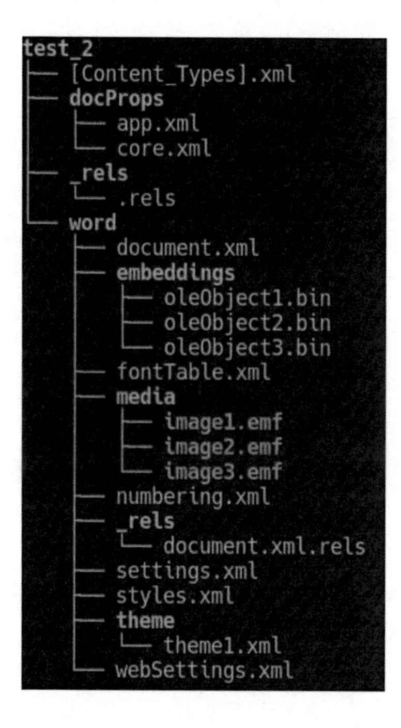

Figure 2-11 Structure of DOCX file

The internal structure of the DOCX file format is shown in figure 2-11. The structure can be interpreted as follows:

- **`[Content_Types].xml`**: every DOCX file must have this file in the root of the zipped package. This file lists all the content type, known as media type or MIME type[44], of the media included within the package.
- **`docProps`** folder: containing the whole DOCX package properties and metadata.

- *app.xml*: Extended file properties (e.g., number of pages and words, application name and version)
- *core.xml*: Document properties (e.g., creator name, creation date, last modified date, revision)
- Thumbnail preview if existed.

- **_rels** folder: storing the top-level relationship of the package, identifying the relationship of each part to others of the DOCX package. The "*.rels*" file indicates the location of the file holding the document contents, which generally is "*word/document.xml*".
- **word** folder: the main folder containing all the contents of the DOCX document. The *document.xml* file must exist inside this folder. The details of other files and folders are explained below:

  - **document.xml**: the main XML of the document contents, holding text information and references to other parts such as media, embedding, styles, header, footer
  - **embeddings folder**: all of the embedded objects are saved here.
  - **media folder**: storing media files such as image, sound, videos.
  - ***"word/_rels/document.xml.rels"***: holding the relationship of items (images, videos, sounds, embeddings) that are referenced in the main content "*document.xml*".
  - ***"styles.xml"***: details of styles are used throughout the document.

To extract ChemDraw contents out of DOCX documents, OLE objects need to be retrieved. The figure below illustrates an example of tracing the ChemDraw OLE contents of the DOCX file in figure 2-11

Figure 2-12 OLE objects and images in the DOCX file

Fortunately, DOCX explicitly stores OLE objects in the main content file "`document.xml`". As shown in figure 2-12, we can find the OLE object of ChemDraw and its image are saved inside an XML *object* tag, with the relationship identifier "*rId8*" and "*rId7*" respectively. Using these identifiers, we can locate the data files via the relationship described in "`word/_rels/document.xml.rels`".

In figure 2-12, OLE contents are saved in "`embedding/oleObject2.bin`". This binary file is `ObjectPool` storage that is covered in the previous section and can be checked for its existence in figure 2-11. In addition, the image "`rId7`" can also be found by following the path "`media/image2.emf`", this image is used later by ChemScanner for the preview purpose of the ChemDraw sketches.

2.2.3. ChemDraw XML (CDXML)

CDXML file format XML-based variation of the CDX file format. Therefore, with a structure similar to CDX files, CDXML files consist of

- A fixed header

```
<?xml version="1.0" encoding="UTF-8" ?>
<!DOCTYPE CDXML SYSTEM "http://www.cambridgesoft.com/xml/cdxml.dtd">
```

- The XML formatted of tagged items. These object with their properties are enclosed with "**`<CDXML>`**" XML tag.
- And ending "**`</CDXML>`**" tag

Unlike the bit-level accuracy of CDX format, CDXML is more friendly to whitespace characters (space, tab, newline, carriage-return) as long as the file contents follow XML format.

In CDXML, tagged items, or objects, are represented as XML nodes. The XML node's attributes are the properties of each object. Figure 2-13 shows an example of one CDXML holding information of *ethane*

```
<?xml version="1.0" encoding="UTF-8"?>
<CDXML CreationProgram="ChemDraw JS 2.0.0.4" Name="ACS Document 1996">
    <page id="12" Width="1682" Height="889.33">
        <fragment id="8" Z="2">
            <n id="7" p="751.33 110" Z="1" AS="N" Element="6" />
            <n id="9" p="763.80 102.80" Z="3" AS="N" Element="6" />
            <b id="10" Z="4" B="7" E="9" BS="N"  Order="1" />
        </fragment>
    </page>
</CDXML>
```

Figure 2-13 A simplified CDXML containing an ethane (C2H6) molecule

The above example shows a CDXML file containing the same molecule (*Ethane*) as figure 2-9. The CDXML file can be interpreted as below

- One `Page` with the identifier equal to 12. `Page` object corresponds to one sketch page in the ChemDraw program. The width and height of the page are defined by properties of the same name.

- `Page` object has one `Fragment` child object, has an identifier equal to 8.
- The molecular fragment above has two `Node` objects, represented by "<n>" tags, and one `Bond` object.
    - `Node` object usually refers to a single molecule atom, with atomic numbers expressed by the "`Element`" property. In this case, we have two carbon atoms since both nodes have the atomic number equal to 6.
    - Similarly, Bond objects refer to molecular bonds, with the bond order indicated by the "`Order`" property. The bond endpoints are determined by "`B`" and "`E`" properties, stand for atoms at the beginning, and the ending of the bond, respectively. In this example, two atoms above are connected by one single `Bond`, which has an ID of 10.

By retrieving and parsing molecular structural contents and graphics from CDX and CDXML files, ChemDraw mining can overcome most of the challenges in the optical chemical structure recognition (OCSR), not only the accuracy of molecular structure but also the processing time duration. However, because of the limited existence of ChemDraw files or documents that embed ChemDraw schemes, the research on the OCSR system is gaining more attention in the development of the chemical information retrieval system.

In the next part, we will introduce and discuss some current state-of-the-art approaches to chemical information retrieval.

2.3 Literatures Review

As described in the previous chapter, the chemical information retrieval system consists of two major components: chemical entity recognition (CER) and linking detected entities with their corresponding structural representation in images throughout the document. In this section, we will introduce a brief overview of current state-of-the-art research in chemical entity recognition and chemical structure extraction from images.

2.3.1. Chemical Entity Recognition

In general, there are three general strategies in chemical entity recognition:

- **Chemical dictionary look-up approaches**: the most straightforward approach, by building a dictionary (a set of words) and a matching method. Each paragraph in the document then is compared with words from the dictionary using the matching method to extract the entity. This approach is sufficient for the specific application field or can be used to detect limited chemical names such as trivial names, abbreviations, or brand names.
- **Rule-based approaches**: chemical entities are detected via a set of rules, which attempt to generalize chemical names. Approaches in this kind usually comprised of two primary components: a collection of rules and a rule-matching engine. Rules are commonly made by regular expressions that are manually encoded by human experts to define appropriates rules.
- **Training set based approaches (machine learning):** with the increasing of machine learning (ML) applications in many fields, many research projects have been using machine learning in chemical entity recognition. In machine learning, chemical entity recognition can be considered as:
    - *Classification problem*: the machine learning model is trained to mark individual words within a document with the predefined entity type.
    - *Sequencing-labeling problem*: similar to a classification problem; however, the output label is marked to each word considering the context in a whole sentence. The marked label would be depended on nearby words or elements.

Various of machine learning algorithms have been tested with chemical entity recognition problems, such as decision trees (DT)[45], random forests (RFs)[46], maximum-entropy Markov models (MEMM)[47], and conditional random fields (CRFs)[48–53], which is considered current state-of-the-art method for sequence-labeling and named entity recognition.

In fact, most of the modern state-of-the-art chemical entity recognition systems combine many approaches at different steps. LeadMine[54] uses chemical names from dictionaries as main components for rules generating. ChemSpot[10] applies the CRFs model for IUPAC systematic names and dictionaries to discover trivial names, and abbreviations. ChemDataExtractor[55] trains a machine learning CRF model on the CHEMDNER corpus[56] then integrates dictionaries and rules for the recognition of chemical formulas and database identifiers.

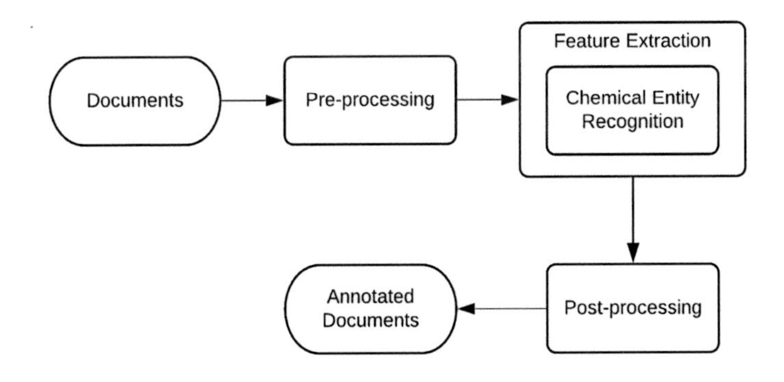

Figure 2-14 General Chemical Entity Recognition flow

Although these approaches combine multiple techniques, they share the same principle steps that are described in figure 2-14. There are three main steps:

- **Pre-processing**: This step usually consists of sentence splitting, text normalization, and tokenization.
  - The normalization process duty is to clean the text, remove special characters, and non-printing Unicode characters.
  - The most crucial step in natural language processing is word tokenization; the process splits text sentences into tokens (e.g., words, compound words, numbers, punctuations). Output tokens then become the input of the next phase of the pipeline.

- **Feature extraction**: the primary step that responsible for chemical entity extraction. There are various phases on this step depend on each system, dictionary-based, rule-based, or machine learning approaches are usually combined together in this part. This step could be divided into two main components
  - **Tagging**: usually of part-of-speech (POS) tagging, assign defined features (e.g., noun, verb, adjective) to words and tokens.
  - **Chemical Named Entity Recognition**: detect and recognize tokens, or entities, and assign it to a specified type. For example,

ChemicalTagger[57] uses OSCAR-Tagger[47] in the first step to identify chemical names, reaction names, chemical prefixes, and adjectives. Then re-scan again using regular expression tagger and part-of-speech tagger to detect chemistry-related terms (e.g., physical state, units) that have not been identified. Machine learning training can be part of this phase by using tagged tokens as training data.

- **Post-processing**: in most cases, this step combines output from *feature extraction* phases or uses an independent solution such as rule-based or dictionary approaches to build the final annotated output.

2.3.2. Optical Chemical Structure Recognition

As introduced in chapter 1, many Optical Chemical Structure Recognition (OCSR) automated software has been published in order to convert images containing structural information into a computer-readable format.

- *Chemical Literature Data Extraction (CLiDE)* [17] employs image processing and artificial intelligence techniques.
- *chemoCR*[18] uses expert rules and supervised learning methods.
- *Imago OCR*[19] uses image processing and dictionaries for abbreviation expansion.
- *OSRA*[14] makes use of image processing methods, character recognition, and confidence estimation.

Although different algorithms are employed, the following steps are common steps between them

- **Image segmentation**: separate text and graphical sections.
- **Vectorization**: the detection of geometry primitives (e.g., characters, lines, segments, triangles).
- **Molecular analysis**: analysis of related molecular elements
  - Atomic labels and charge recognition.
  - Bond recognition, including special bond handling.
- **Post-processing**: validation check, particular case handling such as crossing bonds, aromatic circle.

- **Compilation**: Analysis of graphical regions and reconstruction of the connection table compilation.

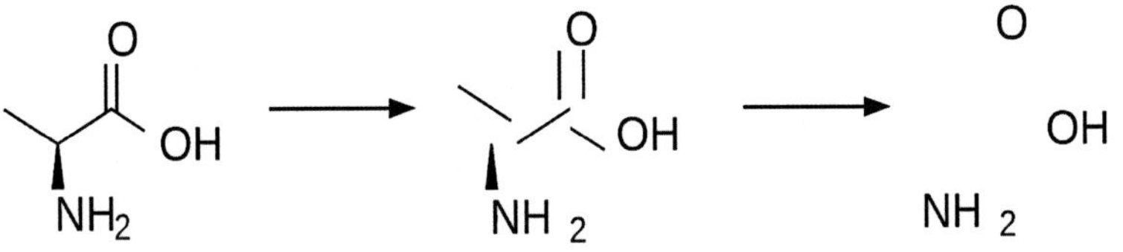

Figure 2-15 Simplified OCSR process of Alanine

The *image segmentation* includes the separation of areas containing chemical structures with others. Segmentation algorithms are varied based on the system. For example, OSRA[14] calculates the rectangular area based on these criteria

- The ratio of the black pixels to the total area is between 0.0 and 0.2
- Aspect ratio is between 0.2 and 5.0
- The rectangle does not intersect with existing structure-containing rectangles
- The width and height are above the minimum values (currently 50 pixels) if the resolution is above 150 dpi.
- The width and height scaled to a resolution of 300 dpi is below a maximum value of 1000 pixels (if the resolution is above 150 dpi)

The isolated chemical structure, as the first image of figure 2-15, then is applied with another *vectorization* algorithm is to classify the graphic (bonds) and text components, like the second and third images of figure 2-15. Each software system has different approaches at this step. For example, OSRA uses the Potrace library[58] to find the position of atoms and bonds.

The *molecular analysis* step is applied to the separated lines and text. Recognized characters are used for atomic label detection. The special characters "+" and "-" are assigned to the nearest atomic label if they existed. The bond detection process includes the identification of the length, position, and direction of the lines to recognize the types of the bonds: single, double, triple, wedged, dotted, dashed, and dashed-wedged. The atoms and bonds recognition may vary on different processing algorithms. During the detection of atoms and bonds, it might occur some particular challenges, as illustrated in figure 2-16. The methods of handling these scenarios take place in the next step.

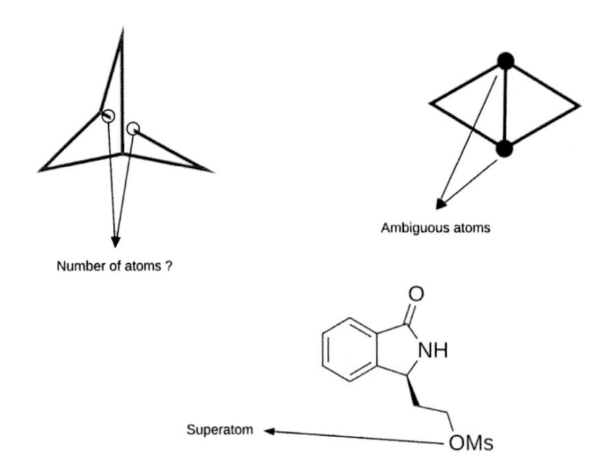

Figure 2-16 Illustrations of some OCSR challenges

Dictionaries of superatoms and abbreviations might also be employed for interpreting the recognized text within schemes in the *post-processing* step. There are various ambiguous schemes, some are illustrated in figure 2-16. Different systems employ their own different approaches. OSRA uses a rule-based approach to manually solve these cases, while chemoCR chooses machine learning for a better scale-up but would need more scheme input data for the training.

The *compilation* process links recognized atoms and bonds above to construct chemical structures. Detected atomic symbols are assigned to each corresponding node of the connection tables, detected lines are to the bond orders. Finally, the constructed connection table is converted into molecular representing such as Molfile or SMILES strings.

2.3.3. ChemDraw Mining

Although current OCSR systems already handle a good amount of chemical diagrams, there are a lot of remaining challenges to solve. For example, figure 2-17 provides an example of the wrong recognition of the most popular open-source solution OSRA

- One wrong carbon connecting to an R-group atom.

- Charges are not detected using OSRA.
- "*BF4*" is ignored

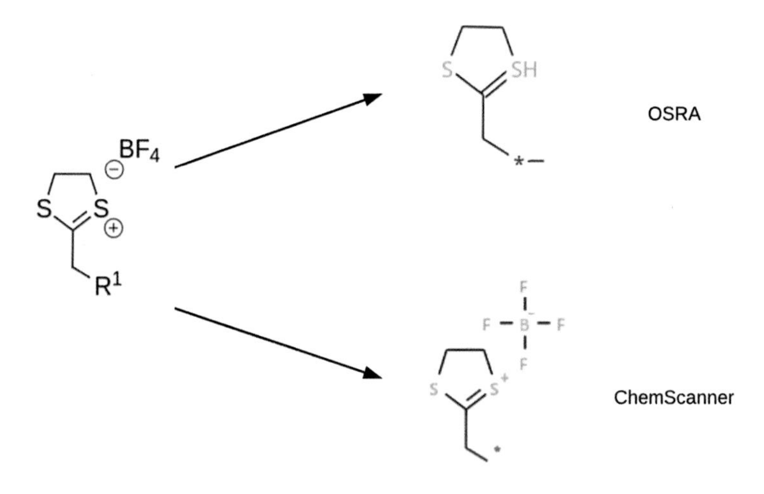

Figure 2-17 An example of ChemDraw sketch processed by OSRA and ChemScanner

In fact, many ChemDraw mining approaches have been developed. InfoChem mentioned about a "CDX Work-up" approach in 2011[59]. In 2012, they integrated CDX processing into the chemistry workflow of Wiley[20], then introduced the *ICSCHEMEPROCESSOR* software package one year later[60]. InfoChem employed a ruled-based combine with the template-based approach in order to resolve ChemDraw scheme interpreting challenges, which they called "optical illusion". NextMove Software also published their work on ChemDraw sketches mining in 2016[22]. Both systems attempt to overcome current OCSR challenges by interpreting ChemDraw CDX file formats.

InfoChem and NextMove Software addressed the limitations of current OCSR systems, such as wavy bonds and crossing bonds detection, brackets interpretation, or unresolved R-group labels. Their works concentrate on molecular structures recognition with limited research on reaction recognition. R-group label interpretation is focusing on text-mining within documents, skipping text information within ChemDraw schemes. Furthermore, both systems are commercial solutions.

Chapter 3. ChemScanner

ChemScanner is a software library developed by Ruby[61], as a gem of RubyGems[62]. *ChemScanner* is developed as a rule-based library for chemical sketches mining, aim to retrieve and interpret chemical information from schemes that are drawn by the ChemDraw software.

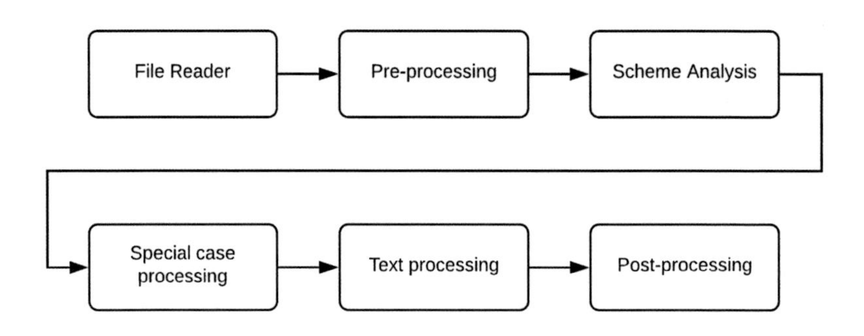

Figure 3-1 ChemScanner overall architecture

Figure 3-1 shows the overall architecture of the ChemScanner. The library can be broken down into six main components

- **File Reader**: reader and extractor of the ChemDraw contents. ChemScanner supports CDX, CDXML files. It also can extract CDX contents from Microsoft Office DOC and DOCX files, or read the CDXML from XML-based file format of the electronic lab notebook from Perkin Elmer[63], the vendor of ChemDraw.
- **Pre-processing**: the component is a rule-based processing component that handles misuse of ChemDraw's internal structure. Due to the basic structure of ChemDraw, some objects are nested within others with ChemDraw intervention that does not fit with expected behavior, this component analyses these situations the categorized them into proper objects groups.
- **Scheme Analysis**: after the cleaning process from previous components, the molecules and reactions are built in this component. Reactions are assembled based on the positions of molecules and graphics that have already recognized from *File Reader* and *Pre-processing* steps.

- **Special case processing**: unusual reactions schemes would be detected and appropriately interpreted with this component.
- **Text Processing**: This component uses curated dictionaries for the detection of abbreviations and superatoms (molecular fragments that are represented by text).
- **Post-processing**: labels that represented molecules across the scheme are replaced by corresponding molecules. The conditions of reactions (time, temperature, yield) would be recognized here. The component uses the interpreted text result from the previous component to generate new molecules and reactions based on the textual information.

In the next section, each of the above components is covered in details

3.1 File Reader

ChemScanner's primary purpose is to read ChemDraw files, including CDX and CDXML file formats. However, the CDX and CDXML files only exist in some repositories such as the US patents USPTO or in-house documents.

Therefore, ChemScanner also has the feature to extract the embedded schemes of ChemDraw within the popular Microsoft Office Word documents that are widely used by researchers for publications. Furthermore, the electronic lab notebook (ELN) published by PerkinElmer, the vendor of ChemDraw, also includes CDXML within the XML-based export file formats of the ELN. By supporting these file formats, ChemScanner can handle various sources of chemical documents, not only the USPTO but also manuscript of publications, theses, or dissertations.

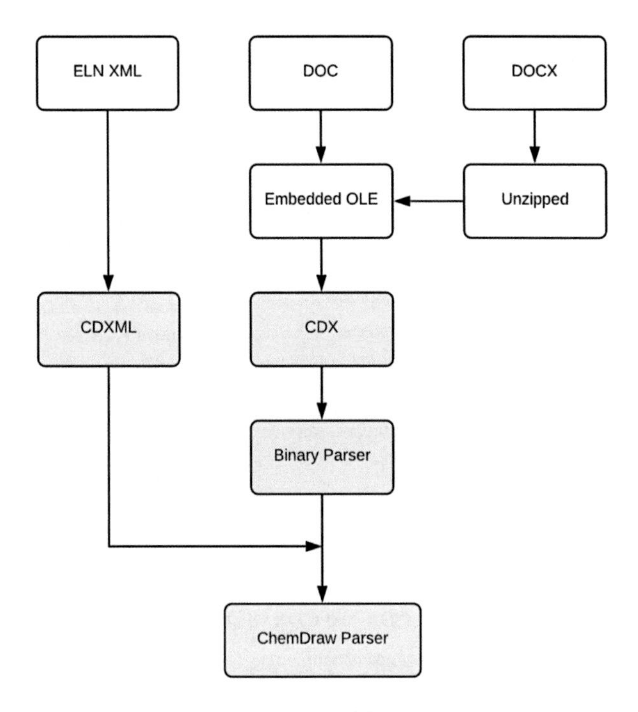

Figure 3-2 Supported file formats handling

ChemScanner supports the ability to read all of these file formats. These formats are processed by *ChemScanner*, as shown in figure 3-2. The reading process of each file format is described below

- **CDXML and XML-based output files from ELN**: These files are XML-based file formats, both of them are human-readable and computer-readable. Parsing XML-based files are straightforward, ChemScanner uses the `Nokogiri`[64] library for the parsing of XML files into nodes and attributes.

- **CDX files**: Unlike XML-based files, CDX files are more complicated to handle. CDX files are parsed using a "`Binary Parser`" to convert binary digits into nodes and attributes.
- **DOCX files**: the DOCX files contain embedded OLE objects inside their "`word/embeddings`" folder, as introduced in chapter 2. The `Nokogiri` library is used for the determination process of ChemDraw's embedded OLE

objects. Once the OLE objects have been retrieved, the `ruby-ole`[65] library is used to read the real CDX contents from them.

- **DOC files**: DOC files are Compound File Binary (CFB) file formats, the `ruby-ole` library is used to read the internal Word binary structure directly. The contents of OLE objects can be extracted, as introduced in chapter 2.

3.1.1. ChemDraw Parser

This is the primary step of the "*File Reader*" component. The contents from the previous steps are parsed so that ChemScanner can "understand" each node and attribute. The parsing process follows the CDX and CDXML specifications published by CambridgeSoft[66]. In this section, the details of the parsing process are covered, which take the ChemDraw's nodes and attributes from the previous step and convert them into *ChemScanner*'s objects.

One critical task of ChemDraw parsing is text retrieval. In ChemDraw, text contents are involved with the following object

- **`Color Table`**: A table that holds the color palettes information that is used throughout the entire document. In CDX format, the "`Color Table`" stores the color information directly as properties of the object. In CDXML format, the color information is saved in the `Color` object, and the table must contain at least one `Color` object.
- **`Color`**: an object holds information of an RGB color. This object only exists in CDXML.
- **`Font Table`**: similar to "`Color Table`", this table contains a list of fonts that are used in the document. In CDX format, the "`Font Table`" stores the font information directly as its properties. In CDXML format, the font information is saved in the `Font` object, and the table must contain at least one `Font` object.
- **`Font`**: an object holds information of font. This object only exists in CDXML.
    - Name of the font
    - The font's character set
    - A unique identifier that is used by other objects when refers to this font

- **`Style`**: holding information of text styling. This object is used only in the CDXML format. In the CDX format, styling information is saved directly within the *`Text`* object.
- **`Text`**: A ChemDraw block for text. The block contains text with styling references that point to the color table or the font table. The text contents may or may not be interpreted chemically. In CDX format, a *`Text`* object must contain a *`kCDXProp_Text`* property. In CDXML format, it must contain at least one "<*s*>" sub-object.

ChemDraw text storing is not as simple as it looks, plain text can be retrieved by parsing only the text object and ignoring its properties. However, since ChemDraw supports text styling and difference styles serve different purposes based on user intention.

For example, users need to use the subscripts to write the chemical sum formula of molecules. A bold number usually represents the label of the molecule, which might be referred later in other places of the scheme or in the document. Retrieving the labels, sometimes called captions, of the molecules would help the linking between molecules and their properties throughout entire documents. The following figure shows an example of how styling text is stored in CDXML.

```
<fonttable>
  <font id="23" charset="utf-8" name="Symbol"/>
  <font id="24" charset="utf-8" name="Arial"/>
  <font id="26" charset="utf-8" name="Georgia"/>
</fonttable>
```

3 α-D-glucose

```
<t id="30" p="637.33 122.66" BoundingBox="637.33 113.97 701.29 127.51" Z="24" LineHeight="auto">
  <s font="24" size="10" color="0" face="1">3</s>
  <s font="26" size="9.9" color="10"> α-D-glucose</s>
</t>
```

4 α-D-glucose

```
<t id="587" p="636.68 160.63" BoundingBox="636.68 151.94 701.04 165.48" Z="27" LineHeight="auto">
  <s font="24" size="10" color="0" face="1">4</s>
  <s font="24" size="10" color="0"/>
  <s font="23" size="10" color="0">a</s>
  <s font="26" size="9.95" color="10">-D-glucose</s>
</t>
```

5 β-D-glucose

```
<t id="29" p="638.66 215.34" BoundingBox="638.66 206.29 703.26 219.91" Z="23" LineHeight="auto">
  <s font="24" size="10" color="0" face="1">5 </s>
  <s font="26" size="10" color="11">β-D-glucose</s>
</t>
```

6 β-D-glucose

```
<t id="1132" p="638.68 253.97" BoundingBox="638.68 245.22 702.44 258.83" Z="32" LineHeight="auto">
  <s font="24" size="10" color="0" face="1">6</s>
  <s font="24" size="10" color="0"/>
  <s font="23" size="10" color="0">b</s>
  <s font="26" size="10" color="11">-D-glucose</s>
</t>
```

Figure 3-3 An example of the text object's structure in CDXML

For the illustration purpose, the CDXML format is used in the example of figure 3-3. In CDX format, the storing of text and styles are slightly different. Nevertheless, the parsing process of both CDX and CDXML are similar.

In figure 3-3, the "*Font Table*" contains three Unicode fonts, that are indicated by the value, "*utf-8*", of the *charset* property. The three fonts are: "Arial", "Georgia", and "Symbol" have their identifiers equal to 24, 26, and 23, respectively. These three fonts are used later in four text objects by referring to their identifier in the value of the *font* property of the *Style* object. In CDXML, each *Text* object must contain at least one *Style* object, represented by the "*<s>*" XML tag. The *Style* object contains the following properties

- font: the ID of font being used.
- size: the size of the font.
- face: a bit-encoded property. There are 7 styling values that are supported by ChemDraw: bold, italic, underline, outline, shadow, subscript, and superscript. These values are sorted from the least significant bit to the most significant bit.

For example, if the text is appeared in bold and underline. The first and third bits are set; others are zero (non-set). The respective face value is $0000101\ (binary) = 5\ (decimal)$

These properties are applied only to the enclosed text. In figure 3-3, all text given in bold has the *face* value equal to 1. In addition, as illustrated, the real saved character of the *alpha* (α) and *beta* (β) character completely different, although their appearances are similar.

In addition, other drawing objects from ChemDraw are also being parsed. The following objects of ChemDraw are parsed by *ChemScanner*.

- **Node**: basic chemical block object of ChemDraw. In most cases, it usually refers to a single atom. However, in some cases, a `Node` can represent other chemical meanings that are indicated by its `NodeType` property.
- **Bond**: the connections between two Node objects, usually refers to a chemical bond.
- **Fragment**: a collection of *Node*s and *Bond*s, usually refers to a molecular fragment or molecule.
- **Geometry**: a geometrical relationship between one or more objects
- **Arrow**: an object that represents an arrow in ChemDraw. It can be a line or arc, optionally with arrowheads on one or both ends.
- **Graphic**: a graphic object, usually a non-chemical object (e.g., a line, circle, or rectangle).
- **BracketedGroup**: represents a set of objects enclosed within brackets, braces, or parentheses.

In general, the parsing process of the listed objects are similar and share many properties in common. In ChemDraw formats, each property value is assigned to a data type, which is specified by the specifications of ChemDraw. Knowing the data type of a property helps to parse the property value correctly. The list of data types that are parsed by ChemScanner is shown below

- **Integer**: includes signed and unsigned integers, these data types are widely used as the values of ChemDraw attributes.
  - In CDXML format, integers are represented as an alphanumeric string.

- In CDX format, integers include signed and unsigned of 8-, 16-, and 32-bit size numbers. As introduced in chapter 2, each property in CDX has the `length` field that indicates the data length. This field is used to determine the size of the data in bytes. An integer has the `length` field's value equal to 1 (byte) must be an 8-bit size number (*INT8* or *UINT8*). Similarly, an integer value stored in 2 or 4 bytes is a 16-bit (*INT16* or *UINT16*) or 32-bit (*INT32* or *UINT32*) size number, respectively.

- **Coordinates**: represents 2D (*2DPosition*) or 3D (*3DPosition*) location of an object. In 2D coordinate spaces, the origin is at the top left corner, and the coordinates increase down and to the right.
    - In CDX format, a *2DPosition* contains an x and a y coordinates as *INT32* values with the order "*y x*", y-coordinate followed by x-coordinate. In contrast with CDX format, a *2DPosition* value in CDXML format is stored differently, "*x y*", x-coordinate followed by y-coordinate.
    - On the other hand, a *3DPosition* in CDX format consists of three *INT32* values that follow the order of "*z y x*", z-coordinate followed by y-coordinate followed by x-coordinate. Similar to 2DPosition, CDXML format has the *3DPosition* order contrast to CDX format, "*x y z*", x-coordinate followed by y-coordinate followed by z-coordinate.

- **Bounding box**: The smallest rectangle that able to cover the graphical representation of the object.
    - In CDX files, *BoundingBox* is stored as four integer values following the order: top, left, bottom, and right edges of the rectangle.
    - In CDXML files, *BoundingBox* is stored as four integer values following the order: left, top, right, and bottom edges of the rectangle.

The figure below shows an illustration of one benzene molecule. The figure contains the CDXML contents of benzene and its depiction. Normally, the bounding box of the object is not displayed. Therefore, figure 3-4 visualizes a bounding box, which is implicitly calculated by ChemDraw, around the benzene molecule. The x-axis and y-axis are also added for showing how coordinates are computed in ChemDraw.

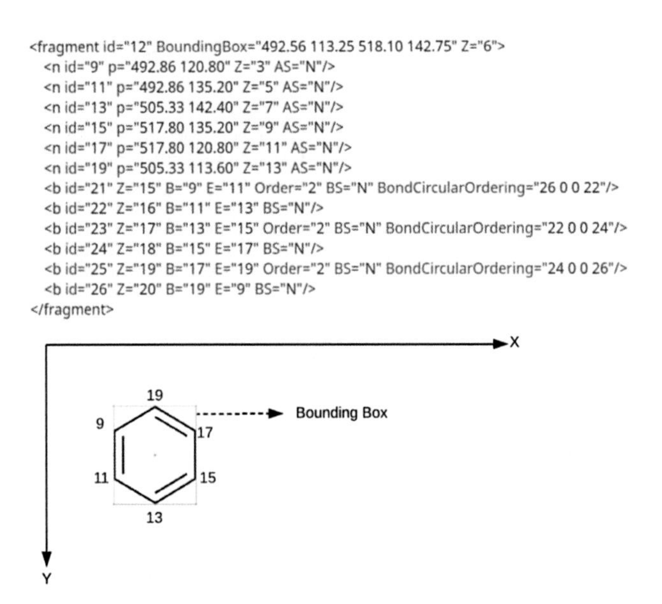

Figure 3-4 Illustration of 2D coordinates of benzene with the bounding box

Each number near each Carbon atom indicates its corresponding identifier of the `Node` object that is shown in the CDXML contents. For example, the Carbon atoms with ID equal to 19 and 13 have the same value of the horizontal location (x-coordinates), which is *`505.33`*; but with the different vertical locations (y-coordinates), that are *`113.60`* and *`142.40`* respectively for ID of 19 and 13. Higher y-coordinates value means it is lower in the displayed picture. Similarly, to the x-coordinates, higher x-coordinates value means it is further on the right in the displayed image.

3.2 Pre-processing

ChemDraw provides many drawing objects for users; however, in some scenarios, these drawing objects are not used exactly as their design purposes. For example, two `Arrow` objects are used to visualize an "X" cross; three or more `Node` objects are placed on the same line so that ChemDraw would display them as a line segment. These scenarios are handled in this section.

3.2.1. Extract nested objects

ChemDraw allows objects can be nested within others, this characteristic allows the critical "`Name To Structure`", or "`Name=Struct`" [67], a feature from ChemDraw, which convert chemical names into structures, can be employed within objects structure. The feature is able to interpret chemical names into the corresponding structures. For example, "`DMF`" would be interpreted as "`O=CN(C)C`" in SMILES format.

Typically, the interpretation process is implicitly executed by ChemDraw whenever a new text has been input to the sketch. The aim of ChemScanner is to retrieve the full structure of molecular structures. Even though ChemScanner has a similar dictionary-based feature to convert chemical names into structures, taking advantage of the "`Name=Struct`" feature to combine them into a hybrid approach would have better results.

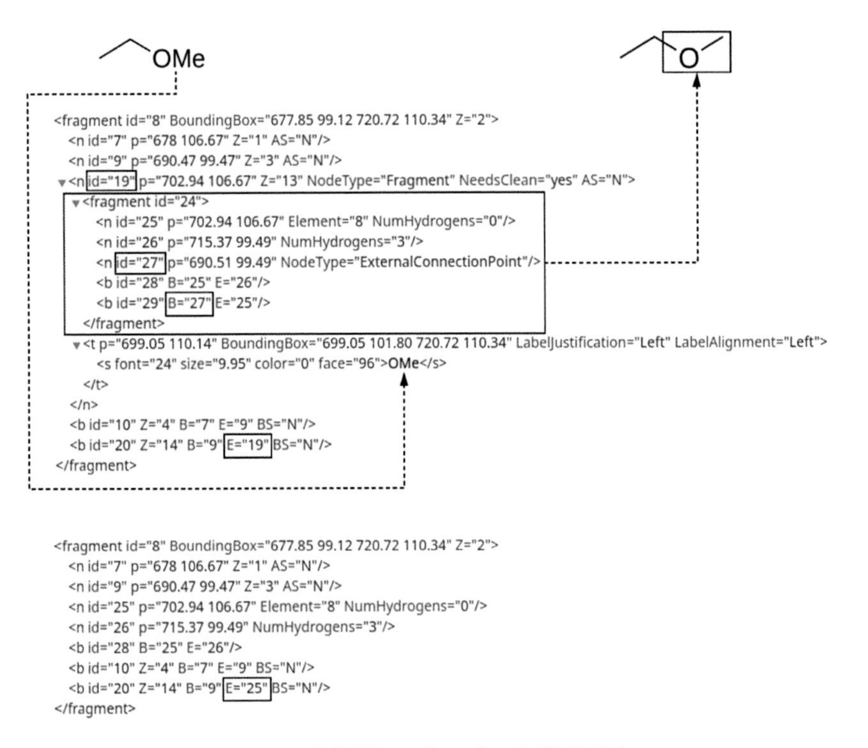

Figure 3-5 "Name=Struct" with CDXML format

The above figure shows an example of how the "`Name=Struct`" is stored in the CDXML format. The `Node` object with ID equal to `19` has the `NodeType` property equal to "`Fragment`", indicates that this `Node` is taking a role as a wrapper for a molecular fragment, not a typical atom. With the node having its `NodeType` equal to "`Fragment`", the Bond object connects to this node is connected to a nested node with `NodeType` equal to `ExternalConnectionPoint`.

The extracted object is also shown in the CDXML format; the ending point of Bond with ID equal to `20` is changed after the extraction. Initially, this bond connects the node with ID equal to `9` to one with ID equal to `19`, which has a nested fragment within. However, since the `Node` object with ID equal to `19` has `Fragment` as the `NodeType`'s value, this `Bond` object's endpoint is another `Node` object with ID equal to `27`.

Therefore, to extract the nested fragment, every connection to the `Node` objects with `ExternalConnectionPoint` and `Fragment` as `NodeType` would be replaced on each other. In the example of figure 3-5, the `Begin` endpoint of the `Bond` object with ID equal to *29* is replaced by the `Begin` endpoint of the `Bond` object with ID equal to *20*. Similarly, the `End` endpoint of the `Bond` object with ID equal to *20* is replaced by the `End` endpoint of the `Bond` object with ID equal to *29* would be replaced with each other. After the transformations, these two `Bond` objects refer to the same connection between two `Node` object with ID equal to *9* and *25*. Therefore, one `Bond` object is omitted, and the `Bond` object with ID equal to *20* remains.

3.2.2. Graphic refinement

Isolated molecules within the rectangle

In figure 3-6, the rectangle is defined as a `Graphic` object which its `GraphicType` value is `Rectangle`. The `Graphic` object requires a `BoundingBox` property, which is covered in section 3.1. Since the object is already a rectangle, the `BoundingBox` property represents itself, and its value consists of two points that are the *left-top* and the *right-bottom* edges of the rectangle.

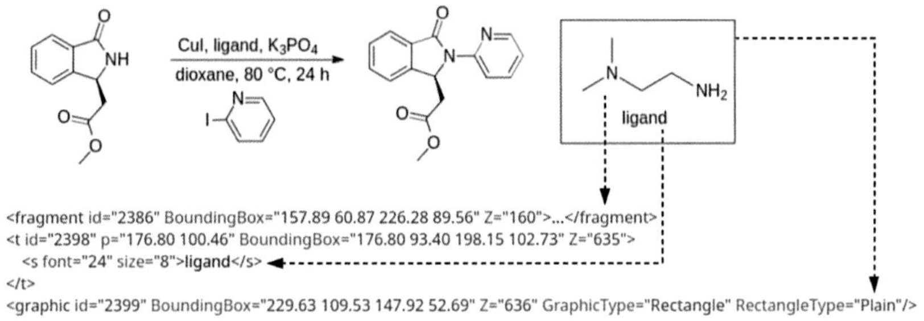

```
<fragment id="2386" BoundingBox="157.89 60.87 226.28 89.56" Z="160">...</fragment>
<t id="2398" p="176.80 100.46" BoundingBox="176.80 93.40 198.15 102.73" Z="635">
  <s font="24" size="8">ligand</s>
</t>
<graphic id="2399" BoundingBox="229.63 109.53 147.92 52.69" Z="636" GraphicType="Rectangle" RectangleType="Plain"/>
```

Figure 3-6 Example of an isolated ligand

Using the `BoundingBox` property of the rectangle, *ChemScanner* checks if an object, including a `Fragment` object, a `Node` object, or a `Text` object, is contained by the rectangle and mark the object as isolated.

Line segment detection

There is only one option in ChemDraw to draw a straight line by using an `Arrow` object without a head or tail. Due to that limited line drawing feature, users might try other alternative solutions to reach their goal.

There are cases that users employ the `Bond` objects to draw line segments since, in ChemDraw, the visualization of the `Bond` object and a line segment look identical. One example of the use of Bond objects as line segments is shown in the figure below.

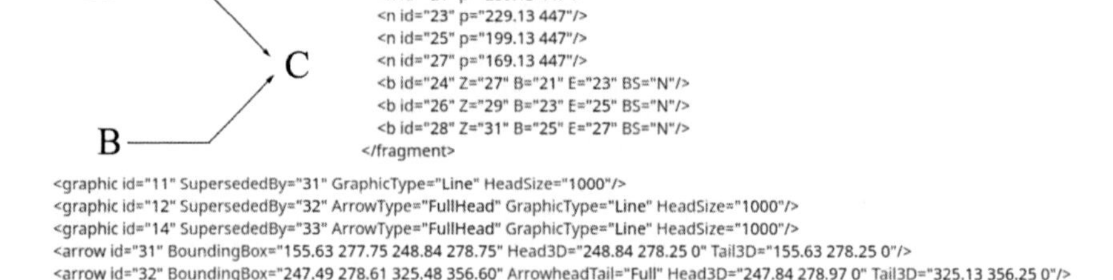

Figure 3-7 Example of line segments

In the example, molecules are represented by letters. Therefore, figure 3-7 describes three molecules `A`, `B`, and `C`. Normally, the figure would be interpreted as two reactions: the first reaction has one reactant A, the second has one reactant B, both of them have C as the product. The first reaction has a one-line segment, which is a "headless" `Arrow` object, and a normal arrow. In the case of the second reaction, it looks similar, a one-line segment and one arrow. This one-line segment is actually a `Fragment` object containing four `Carbon` atoms with three `Bond` objects, as illustrated in figure with the `Fragment` object with ID equal to `22`.

In the current state, ChemScanner detects all the possibility of line segments

- No-head `Arrow` objects
- `Fragment` objects that only have `Carbon` atoms and those `Carbon` atoms are on the same line. The resulting segment takes the first and the last `Node` objects of the `Fragment` as the endpoints.

These detected line segments are used in the next step for a properly arrow interpretation.

3.2.3. Arrow refinement

Segments merging

Unlike ChemDraw, *ChemScanner* defined an arrow as path segments. It means that arrows are comprised of multiple segments, with one final segment containing the arrowhead. The purpose of this section is to interpret ChemDraw's arrows into *ChemScanner* arrows by combining `Arrow` objects with the detected segments from previous steps.

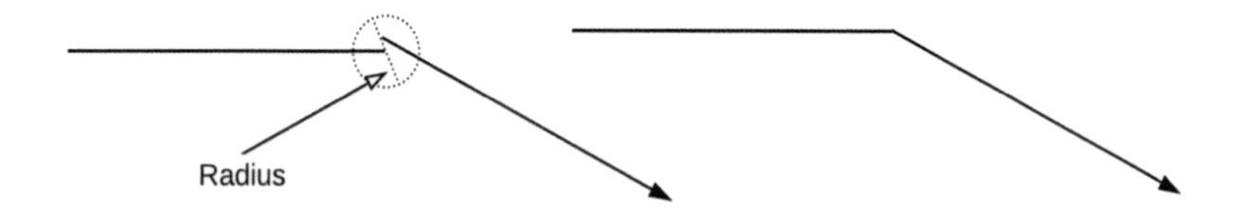

Figure 3-8 Path segments to ChemScanner arrow converting

In order to achieve the converting, the merging of two neighboring segments is needed and is visualized, as shown in figure 3-8. The ChemDraw software does not support to draw multiple consecutive segments so that users have to draw them manually. In many cases, due to the inaccuracy of manually drawing, users would end with the first two segments in figure 3-8. If the endings of the two segments are closed to each other, they are supposed to fit together.

ChemScanner merges the two segments by determining *a radius equal to 10% of the minimum length of the segments*. Using the `BoundingBox` property to retrieve the coordinates of the `Begin` and `End` endpoints of the segments, ChemScanner checks if two segments can be merged together *if the distance of any two points among the endpoints is smaller than the predefined radius*. In figure 3-8, the merged path is illustrated as the right part of the figure.

Defined Patterns

Figure 3-9 describes a list of defined segment patterns that is processed by *ChemScanner*. For illustration purposes, most segments and arrows are drawn in the horizontal and vertical directions. In fact, the refinement process is based on a set of rules that rely on the arrow direction. The patterns shown in figure 3-9 are not only limited to horizontal or vertical directions but can also be applied for paths that have more than two segments in any direction.

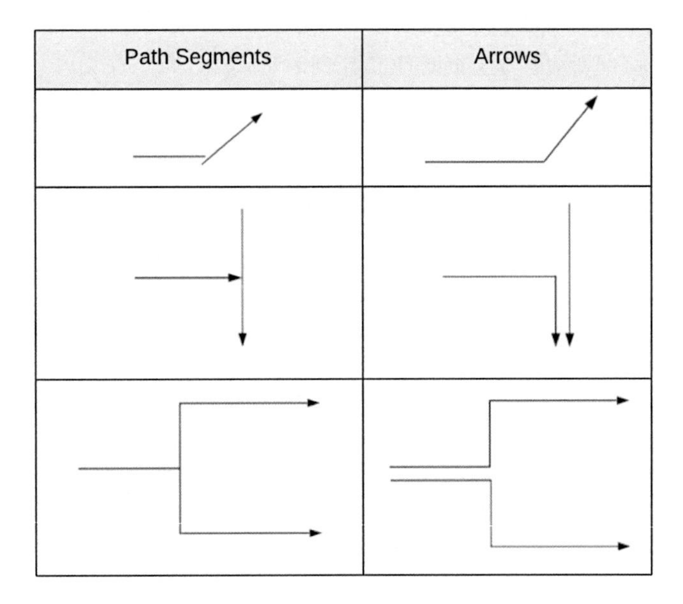

Figure 3-9 Illustrations of the defined segments patterns

Reaction status

In addition to normal arrow, which is generally interpreted as a *successful* reaction, researchers also use different methods to represent a reaction is *failed* or is *planned*. Figure 3-10 summarizes the rules that are used by *ChemScanner* to detect the status of a reaction based on the reaction arrow.

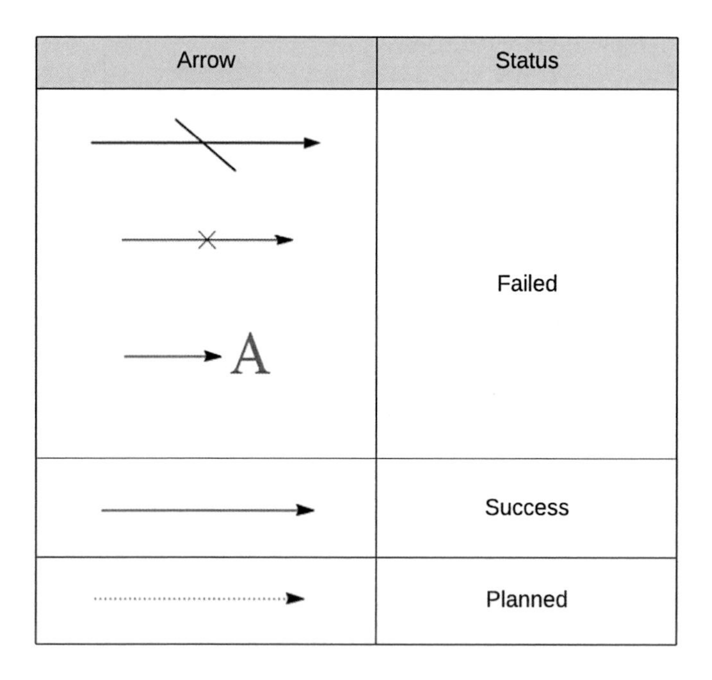

Figure 3-10 Reaction status detection

On the one hand, a successful reaction is represented by a normal solid arrow, and a planned reaction is indicated by a dashed arrow. On the other hand, *ChemScanner* recognizes a failed reaction via three rules:

- The arrow is intersected with one or two segments, as shown in the first case of the *failed* status in figure 3-10.
- The arrow is marked as a crossed directly in the ChemDraw software.
- The products of the reaction are drawn in red color. The reaction assembling is covered in section 3.3.

If a reaction satisfies at least one of the above conditions, the reaction is considered as a failed reaction.

3.2.4. Molecule compilation

RDKit Ruby-binding library

Retrieving molecules and reactions from sketches of ChemDraw is the responsibility of `ChemScanner`. To achieve this goal, `ChemScanner` uses `RDKit`[30], a well-known open-source cheminformatics software, for the construction of atoms and bonds of molecules.

`RDKit` is developed in `C++` and does not have an interface for `Ruby`. The Simplified Wrapper and Interface Generator[68] (*`SWIG`*) is employed to create `rdkit_chem`[69], a Ruby gem for the `RDKit`.

In the `rdkit_chem` library, atoms and bonds are defined as classes with the same name: *`Atom`* class and *`Bond`* class. On the other hand, with molecular fragments, there are two classes for different purposes:

- *`ROMol`* (read-only molecule) responsible for reading molecular information from a fixed molecule topology. Basic molecule formats (e.g., SMILES, Molfile) can be retrieved from this class.
- *`RWMol`* (read-write molecule) responsible for the editing of the current molecule object. *`RWMol`* is inherited from *`ROMol`* so that the SMILES and Molfile of the molecule can also be read using this class. This *`RWMol`* is employed by ChemScanner for the construction of the molecules and also widely-used across the library for molecule manipulation.

Ionic Molecules

In Organic Chemistry, it is common to see molecules that are connected to each other with ionic bonds identical to the below figure.

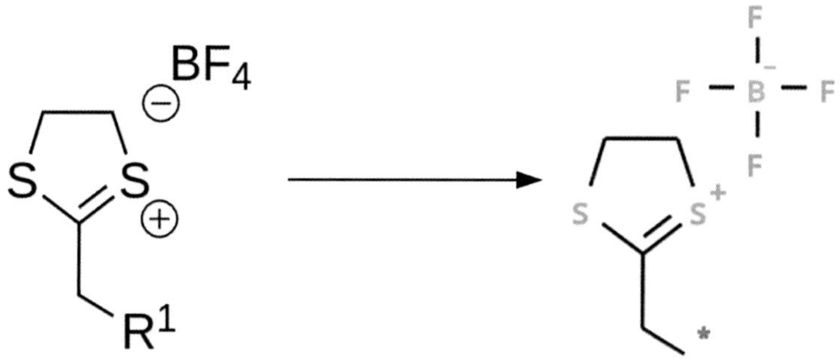

Figure 3-11 Example of ionic molecules detection

`ChemScanner` merges these ionic molecules with the following steps

- Detect molecules with positive and negative charges and group the molecules into two separate lists.
- On every charged molecule in the list above, calculate the distance from the charged *`Atom`* object to the *`Atom`* object of other molecules that have the opposite charge value. For example, if a molecule containing one `Sulfur` atom has a positive charge "*2+*", it is paired with other molecules containing an atom with the negative charge "*2−*".
- Compare the distance D_{mol}, received from the previous step, with a predefined distance D_{min}. If $D_{mol} < D_{min}$ then merge the two respective molecules into one molecule. If $D_{mol} \geq D_{min}$ then the two molecules are not merged together. The comparison is needed in case of users intentionally draw a molecule with charges.

At the current state of development, ChemScanner only supports the merging of molecules having one charged atom.

Fragments to Molecules

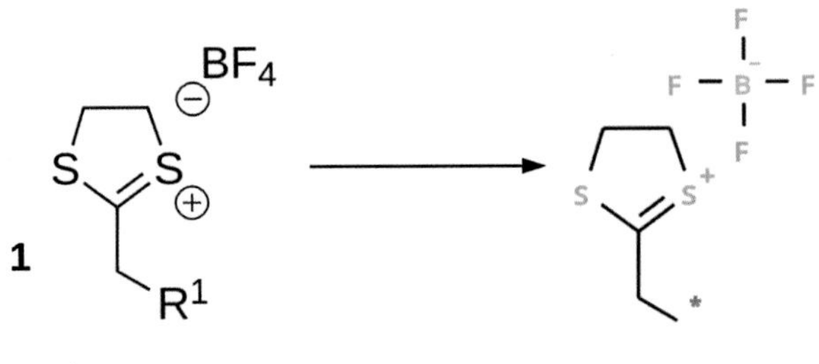

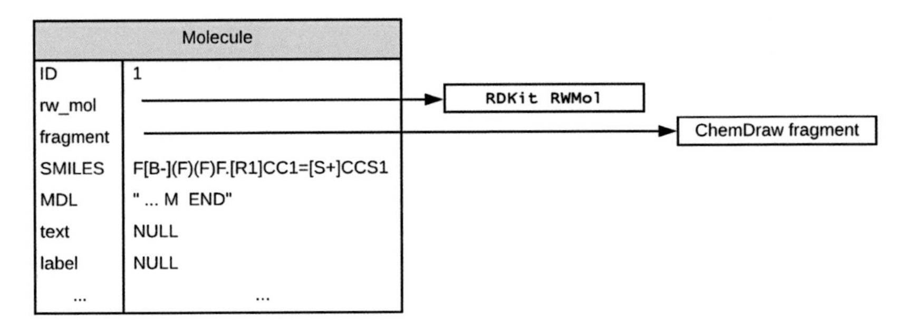

Figure 3-12 The Molecule class of ChemScanner

At this state, *ChemScanner* converts `Fragment` objects into a class of *ChemScanner*, the `Molecule` class. The class contains many properties that are used by *ChemScanner* for further processing. Below are some main properties:

- **`ID`**: An identifier number, it usually takes the value from the `ID` of the `Fragment` object. In case it is a generated molecule object created by *ChemScanner*, *ChemScanner* assigns a new unique value to it. The `ID` should be unique in the file scope.
- **`rw_mol`**: a reference to an RDKit RWMol object.
- **`fragment`**: a reference to parsed `Fragment` object of ChemDraw.
- **`smiles`**: contains the SMILES representation of the molecule.
- **`mdl`**: contains the Molfile representation of the molecule.
- **`text`**: contains the associated text with the molecule
- **`label`**: contains the label, or the caption, of the molecule.

Each `Fragment` object, not include the `Fragment` has marked as a line segment from the 3.2.2 step, is converted to a `Molecule` object. At this point, only the `smiles` and `mdl` properties are populated. The associated `text` and `label` properties are not assigned to the molecules yet.

3.3 Scheme Analysis

3.3.1. ChemScanner Arrow

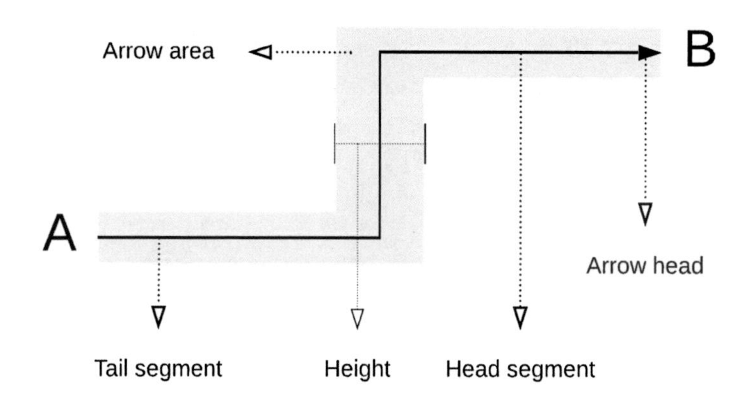

Figure 3-13 Illustration of the arrow segments and arrow area

As introduced in section 3.2.3, `ChemScanner's` arrows are comprised of multiple path segments. `ChemScanner` created a dedicated `Arrow` class to represent the *path segment arrows*. An `Arrow` object can be broken down into the following components, as illustrated in figure 3-13:

- **Arrow area**: a polygon that is made of multiple rectangles following the arrow path defined by the arrow segments. The height of each rectangle varies and can be changed. In fact, its height is changed during the processing of the "*Scheme Analysis*" component.
- **Head segment**: the segment containing the arrowhead.

- **Tail segment**: the segment containing the tail of the arrow. If the arrow contains only one segment, the head segment and tail segment are referring to the same segment.
- **Middle segments**: segments in between the head segment and tail segment, not tail segment nor head segment. There are no middle segments if the arrow contains less than two segments.

3.3.2. Analysis Procedure

The "*Scheme Analysis*" component is responsible for the assembling of reactions in the ChemDraw's schemes. The number of reactions in the scheme is determined by the number of `Arrow` objects, excluding the headless `Arrow` objects that are already marked as line segments. The "*Scheme Analysis*" component consists of two main modules

- **Role detection**: the module is a set of rules that is applied on every detected molecule to determine if the molecule belongs to a reaction. If it does, then determine which role the molecule takes part in the reaction.
- **Arrow area builder**: the module is used to construct the *arrow area* by creating the *arrow area*'s rectangles based on a given height.

The "*Scheme Analysis*" component employs the "*role detection*" and the "*arrow area builder*" modules in the three-steps procedure, which are illustrated in figure 3-14

- **Step 1**: In this step, `ChemScanner` iterates through the molecule and arrow list in the scheme, then executes the following actions on each iteration

    - Build the *arrow area* based on the calculated height. The details about the calculation of height and how the arrow area is constructed is covered in the next section.
    - Apply the *role detection* rules on the molecule to detect the role of the molecule.

- **Step 2**: If the reaction does not have any reagent or solvent, the *arrow area* is rebuilt based on a predefined default value. Otherwise, the *arrow area* of each arrow is rebuilt based on the distance between the reagents or solvents of the reaction and the arrow segments.

- **Step 3**: The *arrow area* that is created from step 2 is kept; the "*role detection*" rules are applied again on every molecule.

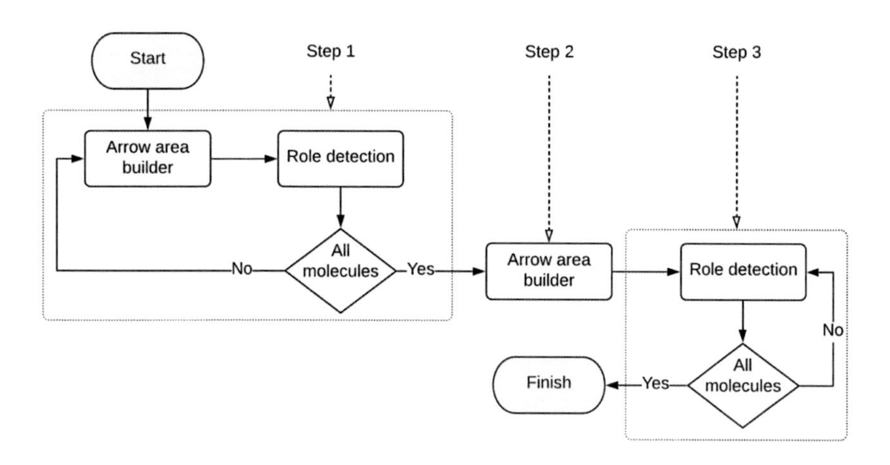

Figure 3-14 The Scheme Analysis procedure

3.3.3. Arrow area builder

As introduced in section 3.3.1, the *arrow area* is the intersection of multiple rectangles with each rectangle is associated with one *arrow segment* so that the *arrow segment* is the midline of the rectangle. The width of the rectangle, in most cases, is equal to the length of the *arrow segment*. The height of the rectangle is the input of the builder module that is used to create the rectangle. If there is no input, the height is assigned a default value that is predefined by `ChemScanner`.

The calculation process of the rectangle's height that was introduced in the first step of the "*analysis procedure*", is described below:

- $distance_{min}$ - the minimum distance between the molecule's bounding box and the arrow segments
- $height_{bbox}$ - the height of the bounding box of the molecule
- The value of the rectangle's height is the sum of the two above values:

$$h = (distance + height_{bbox})$$

Considering an example of one scheme, that is illustrated in figure 3-15, having a two-segments arrow with four molecules A, B, C, and D. Figure 3-15a describes how the *arrow area* is created for molecule C, and figure 3-15b describes the *arrow area* construction process for molecule D.

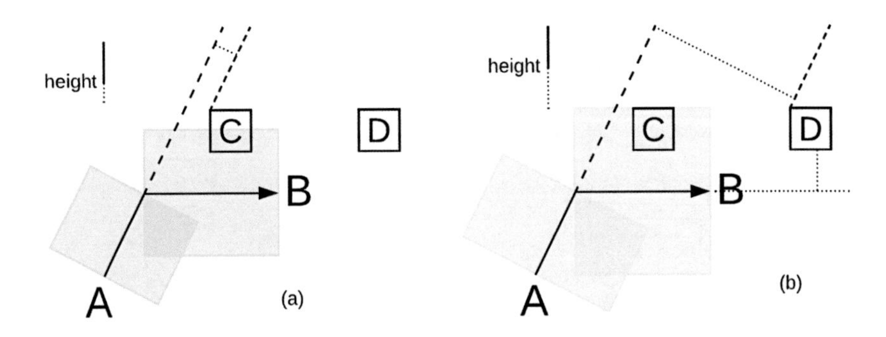

Figure 3-15 Example of the arrow area's construction

In figure 3-15a, the minimum $distance_{min}$ is the distance between molecule C and the tail segment. Based on this value, the height is calculated and shown in the figure, with the length of the dot-line equal to the $distance_{min}$ and the solid line's length equal to the $height_{bbox}$. On the other hand, the minimum $distance_{min}$ is the distance between molecule C and the head segment in figure 3-15b.

The constructed *arrow area* in both cases 3.15a and 3.15b are visualized as grey rectangles in the figure. Based on this arrow area, C belongs to the reaction as a reagent/solvent, while D does not belong to the reaction when the "*role detection*" is applied.

In step 2, the *arrow area* is rebuilt so that the area can contain the whole bounding box of the molecule. In contrast, the height of the rectangle in this situation is calculated by the **maximum** distance between the molecule's bounding box and the arrow segments. For example, the height's value in figure 3-13b is also equal to the maximum distance between the molecule's bounding box and the arrow segments, so that the *arrow* is sufficient to cover the bounding box of molecule C.

3.3.4. Role detection

The "*role detection*" module is made of a set of rules. The ruleset is applied on each detected molecule to determine which reaction it belongs to, and which role it takes part in the reaction.

The detection rules can be summarized as follows

- A molecule is considered as a **reactant** of one reaction if it satisfies all of these conditions:

    a. The molecule or the bounding box of the molecule is on the same side of the **tail** of the arrow.
    b. The line contains the **tail** segment is intersected with the bounding box of the molecule.
    c. The segment, created from the intersection points in `(b)` and the tail of the arrow, is not intersected with any *arrow area*.

- Similarly, a molecule is considered as a **product** of one reaction if all of the following conditions are true:

    a. The molecule or the bounding box of the molecule is on the same side of the **head** of the arrow.
    b. The line contains the **head** segment is intersected with the bounding box of the molecule.
    c. The segment, created from the intersection points in `(b)` and the tail of the arrow, is not intersected with any *arrow area*.

- A molecule is considered as a **reagent** or **solvent** of one reaction if

    a. The molecule is not taking part as a **reactant** or **product** of any other reaction.
    b. The *arrow area* and the bounding box of the molecules tare intersected or contain each other.
    c. If there are multiple results from `(b)`, means that the molecule belongs to multiple *arrow area*, the molecule belongs to the reaction which has

the minimum distance from the bounding box's center to the arrow segments.

- Otherwise, the molecule is considered not to be a part of the reaction.

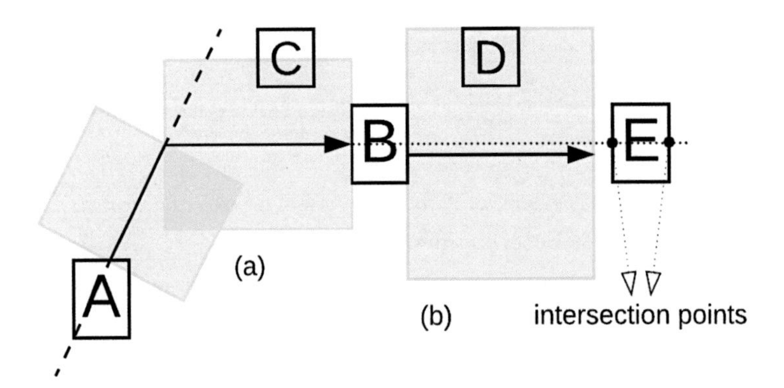

Figure 3-16 An example of role-detection

Considering an example of a scheme containing two arrows and five molecules, as illustrated in figure 3-16. The first arrow *(a)* is a two-segments arrow, while arrow *(b)* is a single segment arrow. By applying the *role-detection* rules on each combination of molecule and arrow, two reactions are obtained as follows

- Reaction *(a)*:
    - Molecule A is a *reactant* of the reaction because the bounding box of molecule A intersects with the line containing the *tail* segment, and is on the same side with the *tail* of the arrow.
    - Molecule B is a *product* of the reaction because the bounding box of molecule B intersects with the line containing the *head* segment, and is on the same side with the *head* of the arrow.
    - The bounding box of molecule E intersects with the line containing the *head* segment. However, the intersection points and the head of the arrow form a segment that intersects the *arrow area* of reaction *(b)*. Therefore, molecule E does not belong to the reaction *(a)*.
    - As introduced in 3.3.3, after the *arrow area* is built, molecule C is considered as a *reagent* or a *solvent* of the reaction.

- Reaction `(b)`:
  - Molecule E is a *product* of the reaction
  - The *tail* segment of the arrow intersects with molecule B's bounding box. In addition, the segment created from the intersection points, and the *tail* of the arrow is not intersected with the *arrow area* of reaction *(a)*. Therefore, molecule B is a *reactant* of reaction *(b)*.
  - Molecule D is a *reagent* or a *solvent* of reaction (b) since it is part of the *arrow area*.

3.4 Special case processing

The detection, identification, and assembling of reaction information from the schemes is a complicated task. Especially when users are trying to embed as much chemical information as possible into the schemes. The *role detection* module is supposed to cover as many different scenarios as possible. However, there are situations that are too complicated to cover for a set of rules set. The *special case processing* component is created to handle these scenarios.

The component consists of multiple modules, each module responsible for processing a specific scenario. In the *role detection* module, a set of rules is applied to every combination of molecule and reaction; the modules of these components are only applied to elements that satisfy specific conditions.

3.4.1. Molecule group

After the *scheme analysis* processing, arrows that only contain reactants or products are *not* considered as *valid* reactions and are removed from the reaction list. Every reaction detected by `ChemScanner` *must* have at least two molecule groups: reactant group and product group. The reactant group or product group is defined as a *molecule group*. Therefore, each reaction contains two molecule groups.

The *molecule group* module is activated on reactions that satisfy the following *conditions*:

i. The reaction has at least one molecule group that has the group size is bigger or equal to 2.

ii. Exists at least one molecule in the molecule group from the condition *(i)* belongs to the molecule group of another reaction.

For each reaction that matches all the *conditions* above, these *actions* are executed

1. For each molecule in the *molecule group* in the condition *(i)*, calculate the $\texttt{distance}_{mol}$ value, which is equal to the minimum of the following distances
    - Distances of the molecule to other molecules in the same group.
    - The distance of the molecule to the head of the arrow.
    - The distance of the molecule to the tail of the arrow.
2. Find the minimum distance, $\texttt{distance}_{min}$, from the list of $\texttt{distance}_{mol}$ from the action *(i)*.
3. For each molecule in the *molecule group* in action *(1)*, use its associated $\texttt{distance}_{mol}$ and the $\texttt{distance}_{min}$ to evaluate the condition: $(\texttt{distance}_{mol} > N * \texttt{distance}_{min})$. *N* is a predefined value by *ChemScanner*, which is currently set to 2.5. If the condition is true, the corresponding molecule is removed from the reaction.

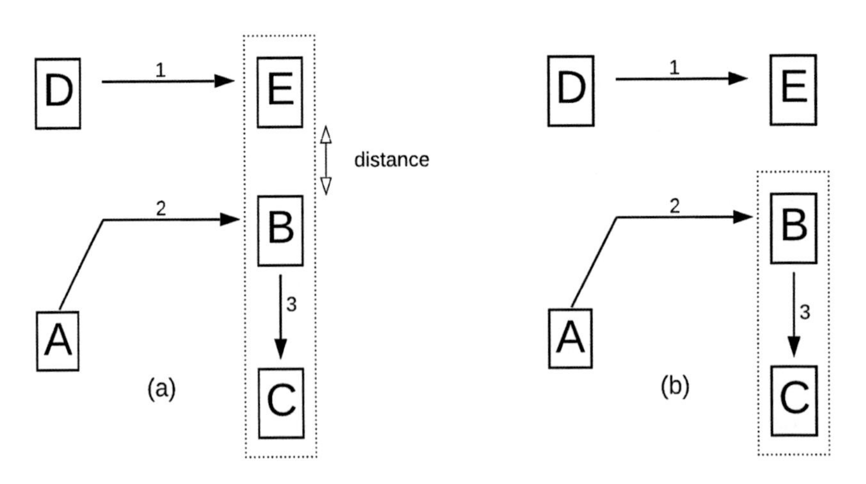

Figure 3-17 Example of the molecule group module

A use case of the *molecule groups* module is illustrated in figure 3-17. After the processing from the *scheme analysis* component, three reactions are obtained. However, by applying the *role detection* module, reaction *3* has two reactants, *B* and *E*, as illustrated in figure 3-17a. Apparently, the result of this detection is not correct.

The reactant group of reaction *3* satisfy the *conditions* from the *molecule groups* module

- The group contains two reactants.
- Molecule E belongs to two reactions: reaction 1 and reaction 3.

If a *molecule group* meets the conditions, the *actions* are executed. In this situation, the value of $\texttt{distance}_{\texttt{min}}$ is the distance between molecule B and the *tail* of arrow 3, the value of $\texttt{distance}_{\texttt{mol}}$ of molecule E is the distance between itself and molecule B. Because the $\texttt{distance}_{\texttt{mol}}$ of molecule E is obviously much bigger than the value of $\texttt{distance}_{\texttt{min}}$, molecule E is removed from the reactant group of reaction 3. In the end, the correct result, as described in figure 3-17b, is obtained.

3.4.2. Multiple line chain reactions

While processing chain reactions, it is common to see molecules in the beginning, or the ending of the line is not drawn when these reactions are spanned across multiple lines. Even though these molecules are not drawn, the scheme is usually interpreted, assuming the molecules exist. The schemes in these scenarios can be categorized into two groups, as illustrated in figure 3-18.

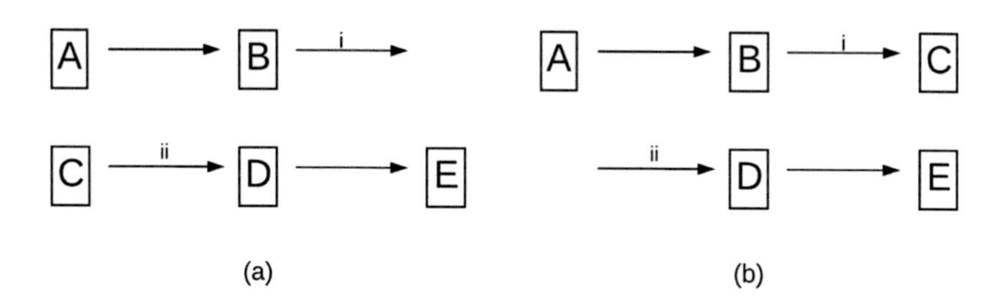

Figure 3-18 Example of common chain reactions

In figure 3-18a, reaction `(i)` is displayed without any product. Nevertheless, molecule `C` is interpreted as a product of reaction `(i)`. Similarly, reaction `(ii)` in figure 3-18b would have molecule `C` as a reactant while, in fact, the molecule `C` is not drawn in the scheme.

This module is responsible for modifying schemes in these scenarios to be more suitable with the original user intention. Unlike the *molecule group* module, this module applies the conditions over the whole scheme rather than a single reaction. To detect if one scheme contains these chaining reactions, `ChemScanner` uses the following conditions

- The scheme contains more than two reactions.
- Every arrow is a single-segment arrow, and the segment is a horizontal line.

If the scheme satisfies the conditions, the following actions are executed to fill the missing molecule in the scheme

- For each arrow, build an arrow area with a small height value, which is predefined in `ChemScanner`.
- Group all the arrows into groups of arrows that are on the same line.
  - Two arrows are considered as being on the same line if the line contains the head segment of one arrow is intersected with the *arrow area* of another arrow, as illustrated in figure 3-19. This comparison is needed due to the manual drawing inaccuracy. Users might place the arrow not precisely on the same *y-coordinates*. The *y-coordinates* of the group is the minimum value of the *y-coordinates* of the arrows within.
- Sort these arrow groups by the *y-coordinates* of each group in ascending order. As explained in chapter 2, bigger *y-coordinates* mean lower position in the scheme. Therefore, the arrow groups are sorted from top to bottom.
- For each group, the arrows are sorted within groups by the *x-coordinates* of the arrow's head, in ascending order. This means that the arrows are sorted from left to right.

- Based on the sorting above, the missing molecules are determined and are filled into the corresponding reactions.

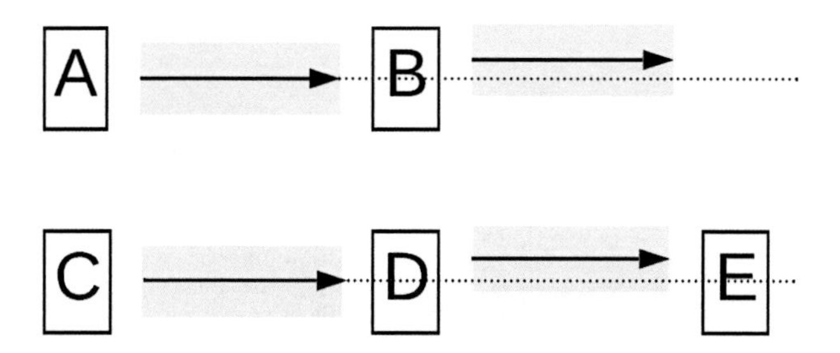

Figure 3-19 Arrows are on the same line determination

3.5 Text processing

In addition to structural information, `ChemScanner` also processes textual information from the schemes of ChemDraw. Textual information in schemes can be categorized into three groups:

- **Abbreviations and superatoms**: text that represents molecular fragments. The text in this category could be a whole molecule (as *abbreviations*), or an alias for a chemical group (as *superatoms*).
- **Molecule text**: text that belongs to molecules. The text holds information about the molecule description and the molecule label.
- **Reaction text**: text that belongs to reactions. The text includes information on the reaction, such as reaction description, reaction time, reaction yield.

3.5.1. Abbreviations and superatoms

As introduced in section 3.2.1, the feature "*`Name=Struct`*" from ChemDraw is able to translate *abbreviations* and *superatoms* into molecules or fragments. However, there are still many *abbreviations* and *superatoms* that "*`Name=Struct`*" is not able to translate. Also, the old schemes cannot benefit from the updating of the "*`Name=Struct`*". To resolve those cases, `ChemScanner` uses a library of 4300 *superatoms* consisting of

- A list of protective groups extracted from the well-known textbook "Greene's Protective Groups in Organic Synthesis" [70].
- A manually collected and curated list of superatoms from the ChemScanner group.

In addition, to complement the abbreviations list, abbreviations, trivial names, and chemical formulas are extracted from the CHEMDNER corpus[71]. The list is curated by using PubChem API[72] to convert them into SMILES. Combining this SMILES list with our internal library, a library is created. This library consists of 6400 abbreviations and chemical names that can be used for the translation of the text to chemical structure.

Furthermore, `ChemScanner` allows users to be able to create their own superatoms and abbreviations library. Once the *user library* is created, `ChemScanner` merges this library with the predefined library. The new library is then used for the translation of *abbreviations* and *superatoms*. In case there is any conflict entry, such as that, an entry exists in the predefined library of `ChemScanner`, in the user library, and also can be translated by the "*`Name=Struct`*" from ChemDraw; the determination of the translated SMILES is based on the priority of the library.

`ChemScanner` prefers the curated list from researchers so that the *user library* has the highest priority; next is the priority of the predefined library, structures from the "*`Name=Struct`*" feature is the lowest. By using this order, if there is any translation mistake from the "*`Name=Struct`*" of ChemDraw or the predefined library of `ChemScanner`, users can easily modify the error by overwriting the translated SMILES to receive the proper interpretation.

3.5.2. Text assignment

At the current state, the *`Text`* objects are only parsed by the *File Reader* component. After the parsing, not only the styled text contents are retrieved but also the coordinates and the bounding box of each *`Text`* object are retrieved. Based on the coordinates and the bounding box of each *`Text`* object, the following actions are executed to determine the association of each *`Text`* object

i. The distance from the *center point* of the `Text`'s bounding box to each molecule is calculated.

    - The distance between a molecule and a point is defined as the minimum distance of the distances from the point to every node of the molecule.

ii. The nearest molecule to the text is determined by the minimum distance, $distance_{mol}$, of the distance list from *(i)*.

iii. If the arrow segments do not contain the projection of the bounding box's center point, the text is not considered as part of the reaction.

iv. In the list of arrows that satisfies the condition *(iii)*, the distance between each arrow and the `Text`'s bounding box is calculated.

v. Determine the nearest arrow to the text by the minimum distance, $distance_{arrow}$, of the distance list from *(iv)*.

vi. Considering the condition: ($distance_{arrow} < N * distance_{mol}$) with N is predefined by `ChemScanner`, by default $N = 2.5$. If the condition is true, the text belongs to its nearest reaction from (v). Otherwise, it belongs to its nearest molecule from *(ii)*.

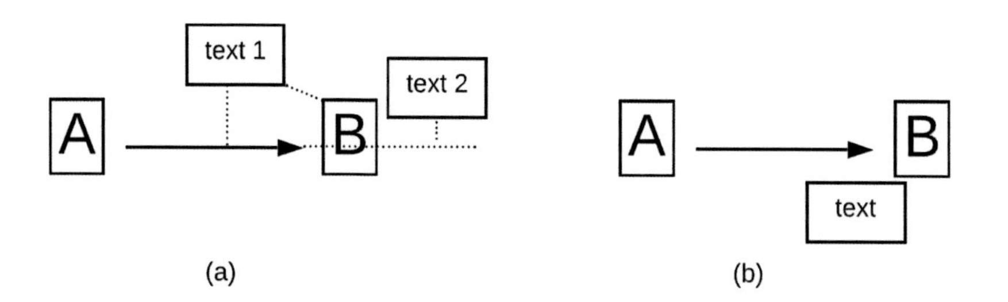

Figure 3-20 Illustration of text assignment

Figure 3-18 describes three examples of the text assignment. In figure 3-18a, the $distance_{arrow}$ and $distance_{mol}$ of the "text 1" object are illustrated with two dotted lines. Approximately, the condition ($distance_{arrow} < 2.5 * distance_{mol}$) is true, so that "text 1" object is considered as textual information of the reaction. On the contrary, the projection point of the "text 2" object's center is obviously not a point in the arrow segment. Therefore, the "text 2" object is not considered as part of the information of the reaction.

Instead, the "text 2" object belongs to molecule B, and its text contents is considered as the description of molecule B.

On the other hand, by applying the actions that are explained above in figure 3-18b, the "text" object would belong to molecule B. However, the scenario is confused even for manual interpretation without knowing the details content of the "text" object.

The situations that cannot be interpreted correctly like this scenario could be edited using the *ChemScanner User Interface* (UI). The *ChemScanner UI* is covered later in chapter 4.

3.5.3. Molecular text processing

Molecule label

Researchers often label a molecule by captioning the molecule by bolded numbers, characters or combine both of them. After the *text assignment* phase, the textual contents and their styling are assigned to corresponding molecules. For each molecule, `ChemScanner` determines its label by reading the bolded text of the molecule.

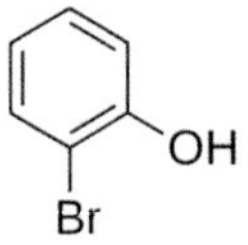

OH (a) 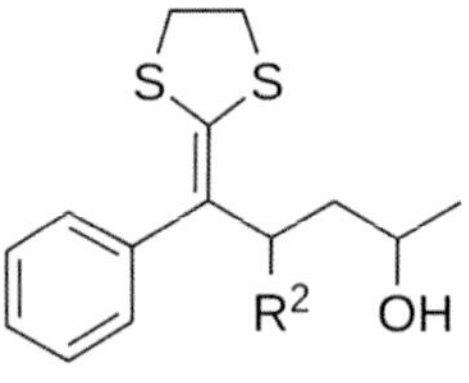
Br
9

S S (b)
R^2 OH
16a: R^2 = H, 94%

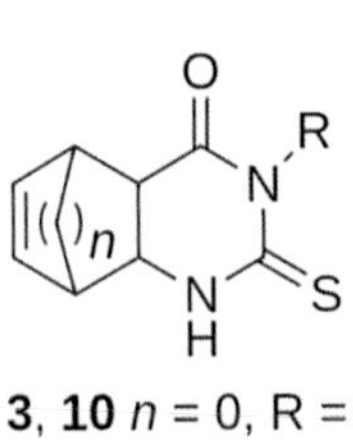

O
R
N
n
N S (c)
H
3, **10** n = 0, R = H
4, **11** n = 1, R = Me

Figure 3-21 Illustration of molecule label

For example, figure 3-21a shows the molecule *2-Bromophenol* that is labeled with a number "`9`". The molecule in figure 3-21b is referred by the label "`16a`"; the text of this molecule also indicates the detail of the Rgroup information.

On the other hand, figure 3-21c illustrates a more complicated example; the bolded text of the molecule contains characters other than the alphanumeric characters. Therefore, the molecule label cannot be detected in the current state. The textual information, in this case, is handled later in the *post-processing* component.

Superatoms expansion

Expanding a *superatom* is the merging process of the molecule that contains the *superatom*, and a molecular fragment that is represented by the *superatom*. In ChemDraw, the *superatom* is represented as a `Text` object that is nested within a `Node` object. As introduced in section 3.2.1, this superatoms expansion process is executed implicitly by ChemDraw for superatoms that are translated by the "`Name=Struct`".

Translated *superatoms* are saved as `Node` objects with the value of the `NodeType` property equal to `Nickname`, or `Fragment`. Besides, if the *superatoms* cannot be translated by ChemDraw, the value of the `NodeType` property is `Unspecified`.

```
<n id="165096" p="136.31 161.92" Z="430" NodeType="Nickname" NeedsClean="yes" AS="N">
 ▾<fragment id="167267">
    <n id="167240" p="136.31 161.92" Z="137"/>
    <n id="167241" p="136.31 173.42" Z="138" NodeType="ExternalConnectionPoint"/>
    <b id="167242" Z="139" B="167241" E="167240"/>
  </fragment>
 ▾<t p="132.98 164.70" BoundingBox="132.98 158.05 144.09 164.85" LabelJustification="Left" LabelAlignment="Left">
    <s font="24" size="8" color="0" face="96">Me</s>
  </t>
</n>
```

```
<n id="55653" p="200.07 86.64" Z="1336" IgnoreWarnings="yes" NodeType="Unspecified" NeedsClean="yes" AS="N">
 ▾<t p="196.96 89.42" BoundingBox="196.96 82.77 216.08 89.57" LabelJustification="Left" LabelAlignment="Left">
    <s font="24" size="8" face="96">OLev</s>
  </t>
</n>
```

Figure 3-22 Superatom in CDXML

An example of two *superatoms* in CDXML format is illustrated in figure 3-22. The first superatom "`Me`", which stands for *Methyl*, is nested within a `Node` object. The *superatom* is already translated by ChemDraw to the Fragment object with ID equal to *167267*. The second superatom "*OLev*" is not recognized by the "*Name=Struct*", so that it is only nested inside the *Node* object with ID equal to *55653*.

All of the *superatoms* listed above are processed by `ChemScanner`. If the library of `ChemScanner` does not have the definition of a *superatom*, translation from ChemDraw is kept. Otherwise, the SMILES from the superatom library is merged with the molecule, as illustrated in figure 3-21.

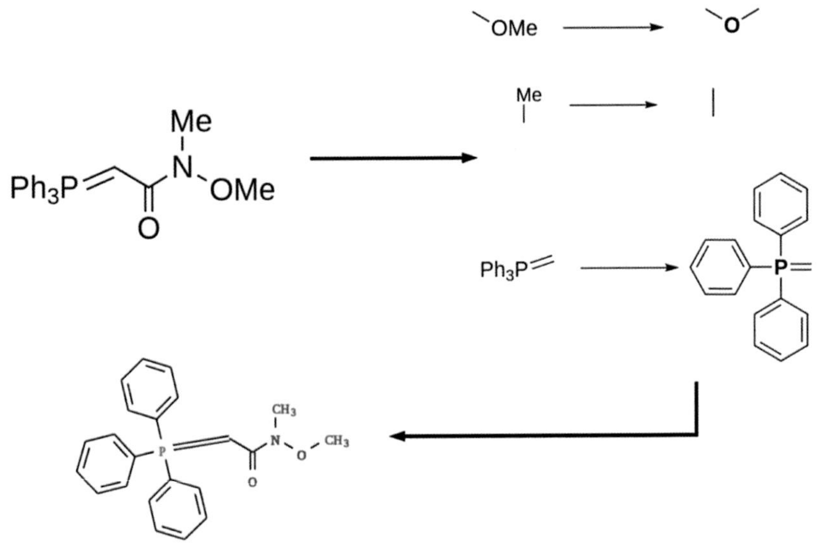

Figure 3-23 Illustration of superatoms expansion

The SMILES in the superatom library of `ChemScanner` is not the canonical SMILES. The SMILES in the library follows an order so that every first character of each SMILES string is considered as the connection point between the molecule and the SMILES structure. It means that the molecule is connected with the SMILES fragment via the first atom in the SMILES string. With the determined connection atom, `ChemScanner` connects the molecule with the SMILES, then uses the `RDKit` library to generate the coordinates for the SMILES structure.

3.5.4. Reaction text processing

The module uses regular expressions to scan for

- **Time**: reaction duration, the unit of reaction time is usually the hour. ChemScanner also able to scan for the day, hour, minute, and second as the time's unit. The time range, such as "`12h ~ 20h`", is also supported.
- **Temperature**: with respect to the reaction temperature, both Celsius and Fahrenheit are detected by *ChemScanner*. Similar to the time property, the temperature range, such as "`20°C ~ 25°C`", is also supported.
- **Yield**: the yield of the reaction. Unlike previous properties, in order to retrieve the yield, ChemScanner not only scans the reaction text but also the text of every product of the reaction.

The abbreviation translation is also processed by this module. Besides the abbreviation library, `ChemScanner` has a separate list consist of 16 solvents that are widely used in organic chemistry. Based on these two lists, every word of the reaction text is used for a lookup process to determine the additional reagents and solvents of each reaction.

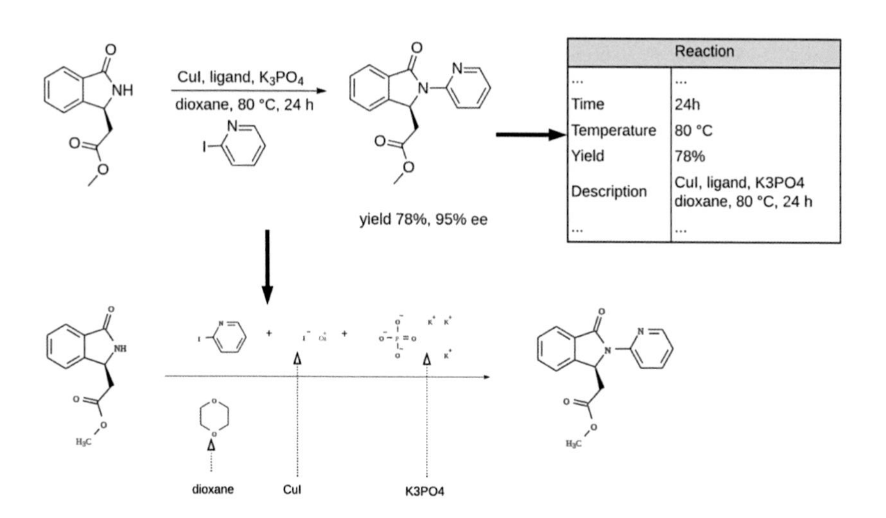

Figure 3-24 Reaction text and abbreviations

An example of abbreviations translation and the detection of reaction information is described in figure 3-24. In this example, "`CuI`", "K_3PO_4", and "`dioxane`" are translated into molecular structures. The reaction temperature and time are extracted from the text within the *arrow area*, while the reaction yield is retrieved from the text of the reaction's product.

3.6 Post-processing

The *special case processing* component is assigned to handle unusual scenarios that can be resolved by using the graphical information as patterns for reaction detection. The *post-processing* component, on the other hand, is responsible for resolving specific situations based on the extracted textual information.

3.6.1. Text as molecules

Reactants or products as abbreviations

Abbreviations are used not only as reagents or solvents but also as reactants or products like illustrated in figure 3-25

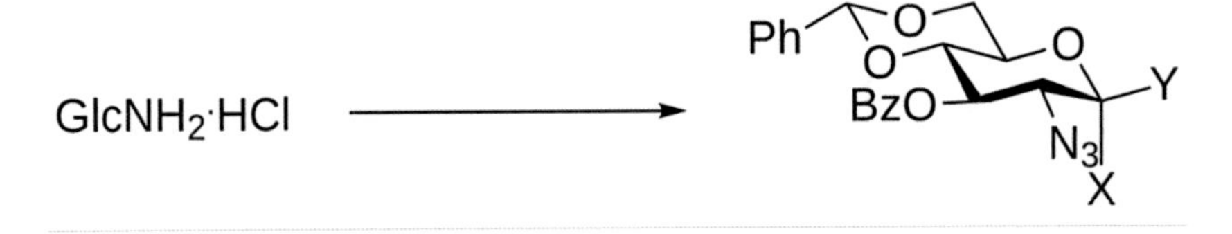

Figure 3-25 An example of text as a reactant

Once abbreviations across the scheme are identified from the previous component, `ChemScanner` applies the *role detection* module again on these text to detect if they are part of any reaction. If an abbreviation belongs to one reaction, its corresponding `Text` object is removed from the text list.

Replacing label by molecules

As introduced in section 3.5.3, molecules are usually referred to via the label of the molecule. This method is widely used by researchers when mentioning to a molecular structure in the textual contents of the document.

Practically, there is another use case of the molecule label; the label of a molecule is used to indicate the molecular structure that is drawn within the same scheme. For example, the bolded text “77”, in figure 3-27, is used to refer to the molecule that is drawn separately within a rectangle. This linking method is widely used when the referred molecular structure is too big or complicated to be drawn within the limited space of the *arrow area*

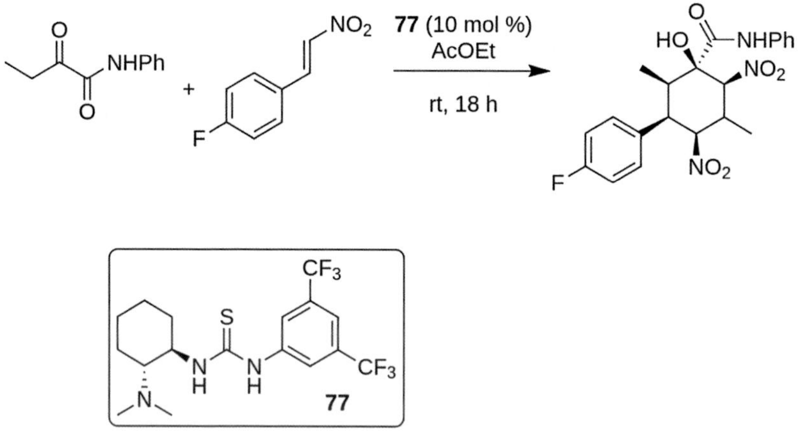

Figure 3-26 Using of molecule label in reaction text

The procedure below is applied to each reaction, in order to replace molecule labels by their corresponding molecules

i. Extract bolded text from the detected reaction text.
ii. Find the existence of a molecule label from the bolded text by comparing each molecule label in the scheme with every word.
iii. If a molecule is found, then add the corresponding molecule into the reagents list of the reaction.

Replacing text by molecules

This step is created to handle scenarios that are similar to situations that are introduced above. Instead of referring to a molecule by its label, the scheme refers to a molecule by or normal word in plain text. For example, figure 3-27 describes a reaction with its catalyst is drawn separately.

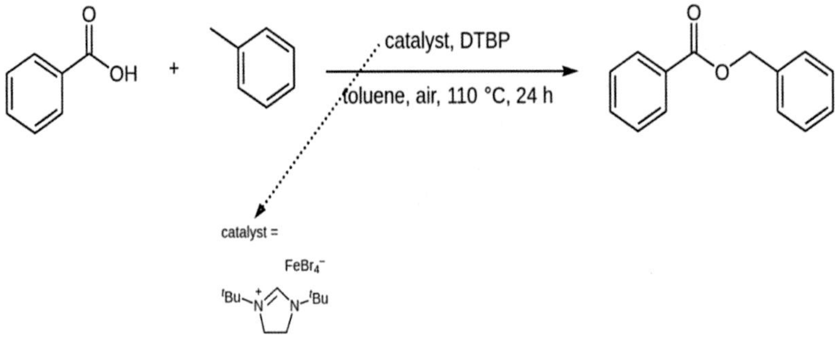

Figure 3-27 Illustration of a text that is used a molecule

The following procedure is applied to each molecule to handle these scenarios

i. Determine the "text representor" of the molecule from the molecule text. If the molecule has a "text representor", then continue to the next step.
   - The "text representor" is defined as a word following by the "=" symbol. For example, "`catalyst =`", "`ligand =`".

ii. Compare the "text representor" with each word of the reaction text.

iii. If a word in the reaction text is equal to the "text representor" of the molecule, then add the corresponding molecule into the reagents list of the reaction.

3.6.2. Multi-step reaction

In addition to describing the reaction description, the reaction text can also be used to describe multi-step reactions, as shown in figure 3-28. The figure describes a two-step reaction via a numbered list. The first step contains a molecular structure which is labeled as "*14*", the second step contains the water molecule.

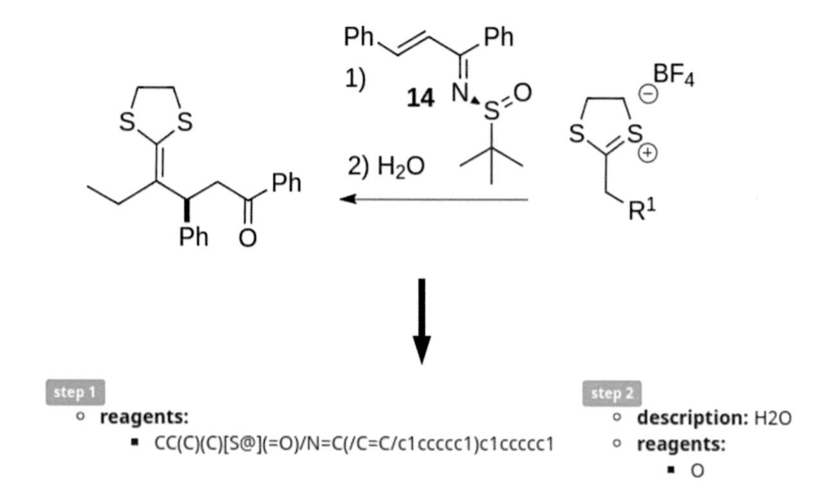

Figure 3-28 Illustration of multistep reaction detection

The result of multi-step detection from the *ChemScanner* processing is shown below the reaction scheme. The SMILES of the molecule with label "`14`" is shown in step 1. Also, the SMILES of water ("`O`") is displayed in step 2.

A reaction is considered as a multi-step reaction if the reaction text satisfies all of the following conditions

- There are more than two lines in the reaction text.
- All lines in the reaction text form a numbered list. The formats of the numbered list must be one of the following
  - 1, 2, 3, …
  - I, II, III, …
  - i, ii, iii, …
  - A, B, C, …

If the reaction is detected as a multi-step reaction, the reaction information (e.g., time, temperature, yield) of each step is processed with the same module "*reaction text processing*", that is already introduced in section 3.5.4.

3.6.3. Generation of new elements

In general, researchers often try to generalize molecules with common fragments into one structure in the scheme, then use textual information to describe multiple molecules from one molecular structure. The same process would also be applied for reactions within a scheme. This section deals with the interpretation of the summarized text from molecules and reactions and generates new elements accordingly.

Molecule generation

`ChemScanner` group molecular information from the molecule text into two types

- **Alias group**: is used to indicate a molecular fragment instead of an atom is a connection with the molecule. *Alias groups* are usually denoted by the symbol *R* in the molecule. Sometimes *X* or *Y* is also employed for the same purpose. When there are multiple *alias groups* in the molecule, the symbol is followed by a number.
- **Repeated group**: is used when the number of atom at a specific node is varied. The target node is enclosed within a brace, a bracket, or a parenthesis and followed by a character as the label of the changing node.

Figure 3-29a illustrates a molecule with one *alias group*, denoted by *R*, and one *repeated group*, denoted by *m*. Figure 3-29b describes a molecule with two *alias group*, that are denoted by *R1* and *R2*.

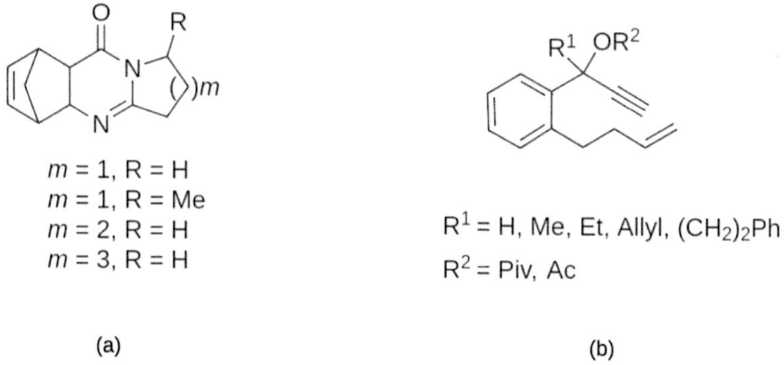

$m = 1$, R = H
$m = 1$, R = Me
$m = 2$, R = H
$m = 3$, R = H

(a)

R^1 = H, Me, Et, Allyl, $(CH_2)_2Ph$
R^2 = Piv, Ac

(b)

Figure 3-29 Illustration of alias groups and repeated groups

For each molecule in the scheme, `ChemScanner` scans the molecule text, and employs regular expression to extract corresponding information of the *alias group* and *repeated group*. Based on the retrieved information, the *molecule generation* procedure is executed on each molecule. The generation process is similar to the *superatoms expansion*, which is already covered in section 3.5.5.

Reaction generation

If a molecule belongs to a reaction, the search domain is not only the molecule text but also the reaction text, and the text contents of other molecules of the same reaction. In addition, with each generated molecule, a new reaction is also generated.

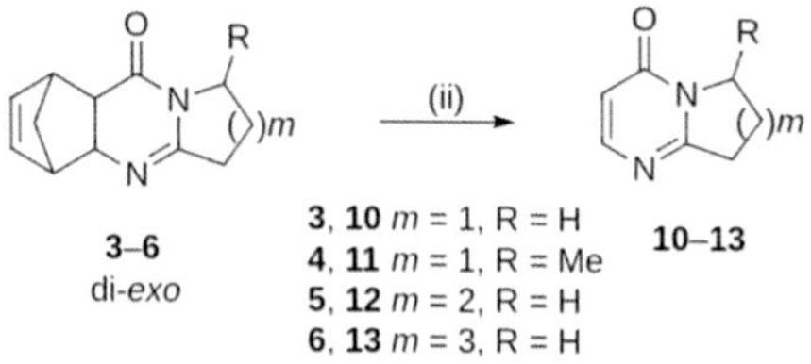

3–6
di-*exo*

(ii)

3, **10** $m = 1$, R = H
4, **11** $m = 1$, R = Me
5, **12** $m = 2$, R = H
6, **13** $m = 3$, R = H

10–13

Reactant label	Product label	Molecule information
3	10	m = 1, R = H
4	11	m = 1, R = Me
5	12	m = 2, R = H
6	13	m = 3, R = H

Figure 3-30 Reaction generation by molecule label

Furthermore, the molecule labels in a reaction can be combined with the *alias group* and *repeated group* information. *ChemScanner* is able to detect these combinations by using regular expressions, then generate new reactants and products with the corresponding labels.

Figure 3-30 illustrates an example of how a label is assigned to a combination of alias group and repeated group. After being processed, ChemScanner generates four reactions. Each reaction has the reactants and products correspond with each record of the table in figure 3-30.

Chapter 4. ChemScanner User Interface

Practically, the chemical information retrieval is a challenging task because of the presence of non-standardized input, especially with the existence of additional textual information. We implemented the *ChemScanner User Interface (UI)* as a computer-aided system for reviewing and editing the scanned output from `ChemScanner`. The system provides a solution for managing multiple uploaded files with different supported file types, the visualization of input data, the reviewing and editing of the results, as well as the storing, re-use, and exporting of chemical data.

4.1 Overview

The *ChemScanner UI* was built on Ruby on Rails[73] framework as a web application. The backend of the website is implemented in Ruby[61], in order to use the Ruby gem `ChemScanner` and `rdkit_chem`[69]. PostgreSQL[74] is employed as the database of the application. The frontend interface is developed in Javascript[75] on the React[76] framework. The overall architecture of the ChemScanner UI application is described in figure 4-1.

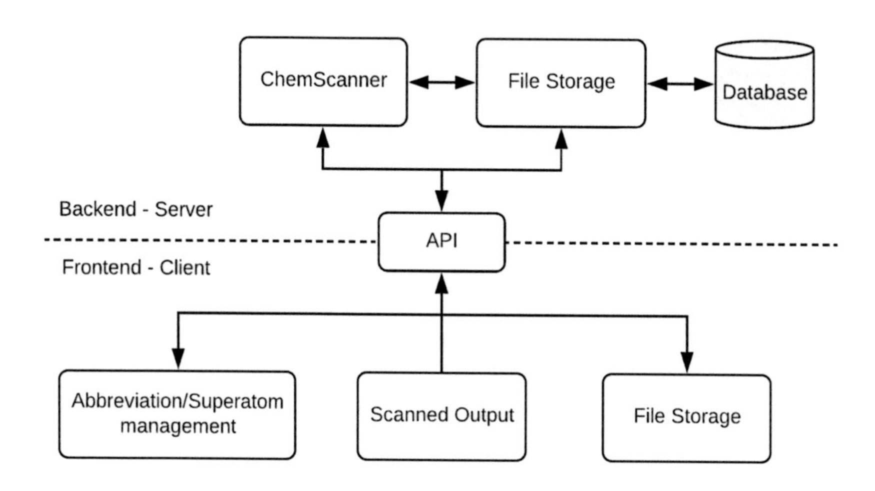

Figure 4-1 Overview of the ChemScanner User Interface

The *ChemScanner UI*, as a typical web application, is made of two parts: the backend and the frontend. The backend is connected to the frontend by using the Application Programming Interface (API). Every action that needed to be performed is sent from the frontend to the server, or the backend, via the API component of the *ChemScanner User Interface*. The backend executes the action and sends the result back to the frontend through the API component.

The backend of the *ChemScanner UI* consists of

- **ChemScanner**: responsible for the extracting of chemical information from ChemDraw schemes and the editing of the library of abbreviations and superatoms.
- **File Storage**: responsible for the management of uploaded ChemDraw files, including
  - Storing uploaded input and output from `ChemScanner` into the database
  - Retrieving uploaded input and its corresponding outputs.
  - Storing and retrieving of the preview image from ChemDraw schemes.

The frontend of the *ChemScanner User Interface* is comprised of three main components:

- **Scanned Output**: responsible for the visualization of the scanned results from the `ChemScanner` library. The editing of the results is also provided by this component.
- **Abbreviation and superatoms management**: the component provides the interface for viewing, creating, and updating of the abbreviation and superatoms library from `ChemScanner`.
- **File Storage**: the respective component to the *File Storage* component in the backend. This component is responsible for displaying and management of uploaded files and their output by communicating with the File Storage component in the backend via the API.

The frontend is built as a single-page application (SPA) that dynamically updates the current page instead of loading entire new pages like traditional websites. The single-page application improves the user experience by avoiding interruption between actions, which is similar to the behavior of the desktop application.

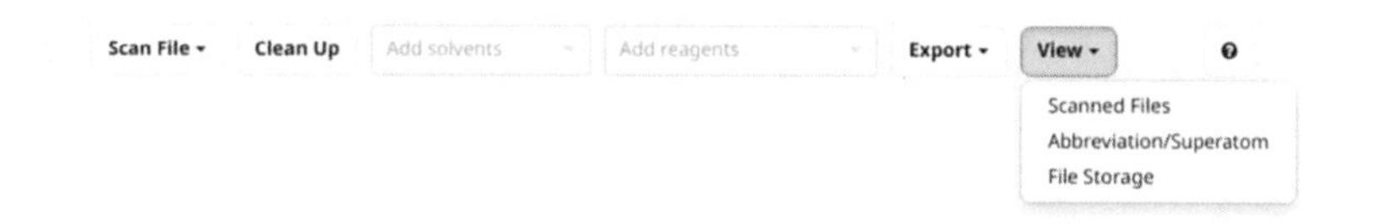

Figure 4-2 The Header Menu of the ChemScanner UI

As a single-page application, the main page of the *ChemScanner UI* is comprised of two modules:

- **Header Menu**: the main menu at the header of the web page, as illustrated in figure 4-2. The menu provides the following options
    - *Scan File*(s): upload a file or multiple files to be processed by the `ChemScanner` library. Users can choose to display molecules or reactions from the scanned results.
    - *Clean Up*: applies the "Clean up" feature of ChemDraw on selected items (Selected items are covered later in section 4.3.1). If the selected list is empty, every item in the "*Scanned Output*" component is cleaned up. Otherwise, only selected items are processed.
    - *Add solvents/reagents*: adds a reagent or solvent from a predefined list into selected reactions. Selected items are covered later in section 4.3.2
    - *Export*: export the outputs from `ChemScanner`. Currently, the `ChemScanner` UI supports the export of data to the Microsoft Excel format and the Chemical Markup Language[77] (CML) format. If the selected list is empty, every item in the "*Scanned Output*" component is exported. Otherwise, only selected items are exported.
- **Main Content**: the contents of three main components which are explained above. The three main components of the frontend are displayed based on the "*View*" option from the *Header Menu*.

The details of the communication between each component of the backend to the frontend are covered in the next section.

4.2 Abbreviation and Superatom management

The library of abbreviations and superatoms is saved internally within the `ChemScanner`. The library can be managed through an interface that is provided by the `ChemScanner` gem. Using this approach, different applications can manage their own library in order to maintain the privacy of the data.

The data flow from the frontend to the backend of the management of abbreviations and superatoms is described in figure 4-3

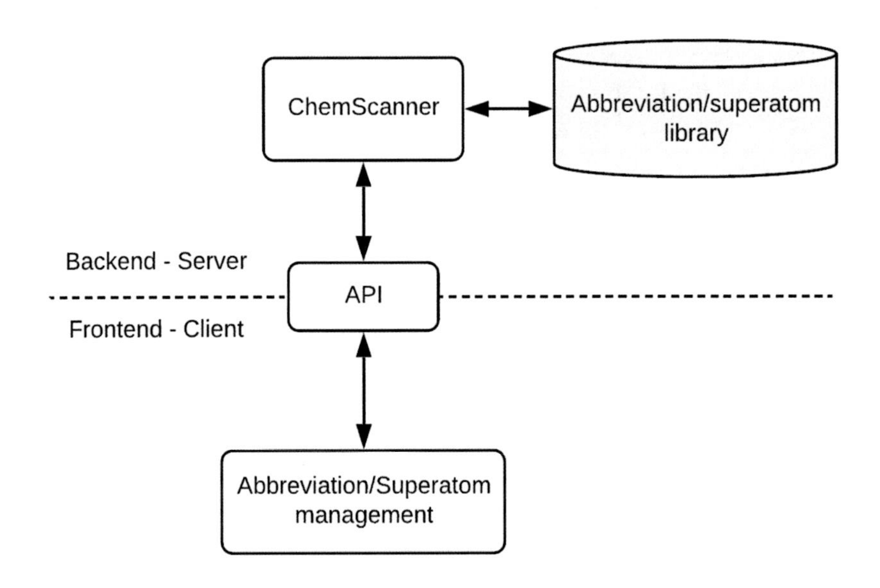

Figure 4-3 Data flow of abbreviations - superatoms management

The `ChemScanner` provides the interface to retrieve the content of the abbreviations and superatoms library, add new entries into the library, and remove existed entries from the library. These actions are executed by users via the "*abbreviations - superatoms management*" component in the frontend.

The "*abbreviations - superatoms management*" component is displayed when the option "*Abbreviation/superatom*" is chosen via the "*View*" dropdown button. The main interface is illustrated in figure 4-4.

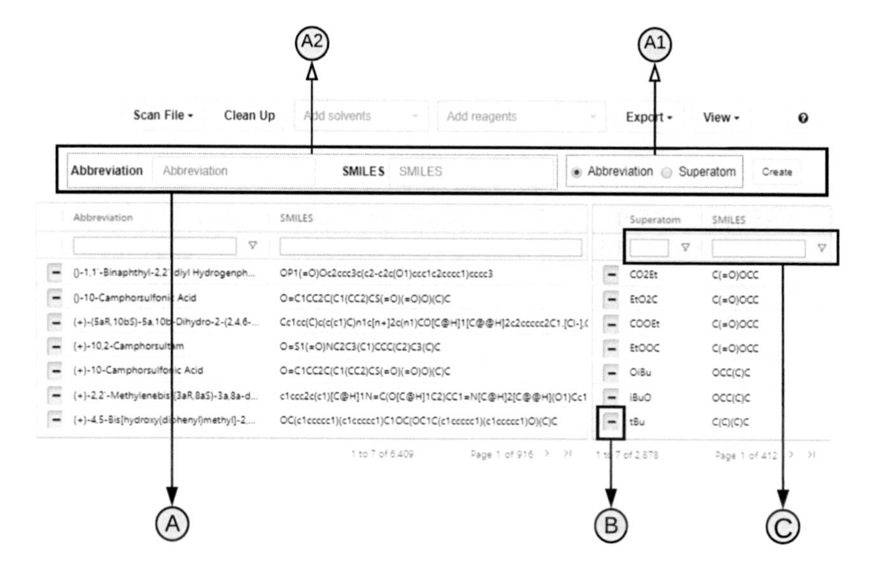

Figure 4-4 The Main interfaces of the abbreviations and superatoms management

The interface of the abbreviations and superatoms management consists of three parts:

- **A**: This section is used to create new entries for the abbreviation list or the superatom list.
    - *A1*: determine the type of the new entry, *abbreviation*, or *superatom*.
    - *A2*: the value of the new entry's name and its respective SMILES.
- **B**: delete the corresponding entry of the superatom list or the abbreviation list. Only entries from the user-defined list can be deleted, entries from the list that are predefined by `ChemScanner` cannot be deleted.
- **C**: filtering, or searching entries within the abbreviation list, or the superatom list. Users can filter by the name or the value of entries.

The abbreviation and superatom lists that are displayed by this page are used in the processing of `ChemScanner`. An example of the using of new abbreviation is described in figure 4-4.

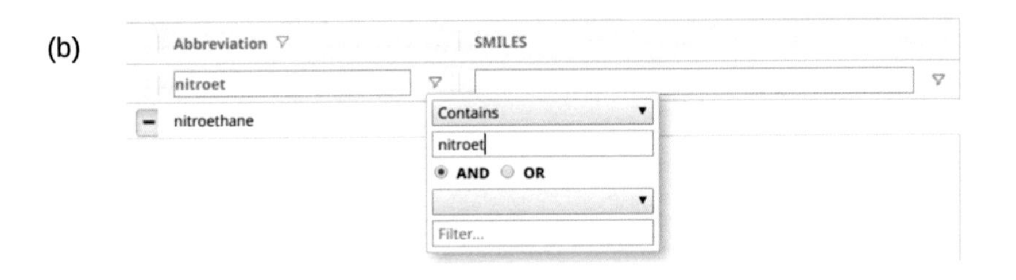

(c)

Figure 4-5 Illustration of add new abbreviation entry

Figure 4-5a shows a simple scheme that is processed by *ChemScanner*. The scheme contains one simple reaction, with `Nitroethane` as the reactant of the reaction. The string "*Nitroethane*" is not recognized via the "`Name=Struct`" feature, and this string does not exist in the abbreviation list of `ChemScanner`. Therefore, the string is not considered as a molecule; the output of `ChemScanner` does not contain any reaction.

A new entry of "*nitroethane*" is added into the abbreviation list of *ChemScanner*, as shown in figure 4-5b. In figure 4-5b, the filtering feature is used to find the "*nitroethane*" entry in the abbreviation list. Due to the existence of the "*nitroethane*" entry, `ChemScanner` converts its SMILES, "`CC[N+](=O)[O-]`", to a molecule and assign the reactant role to the new molecule. The output of the `ChemScanner` is drawn in figure 4-5c.

4.3 Scanned Output

The component is responsible for the management of *ChemScanner*'s results. The input files are uploaded via the "*Scan File*" button in the *Header Menu*. The scanning process from `ChemScanner` is applied to each file in the uploaded files, as described in figure 4-6

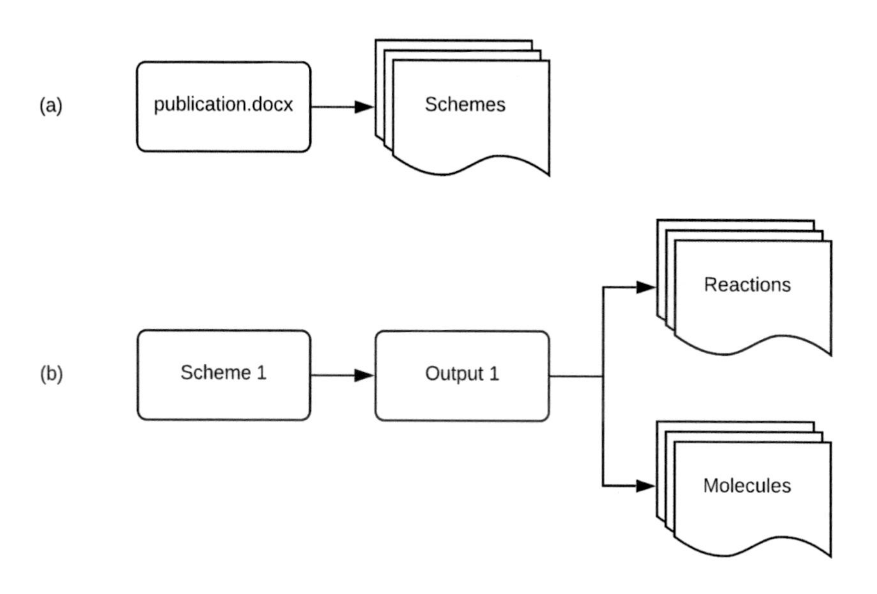

Figure 4-6 The scanning process of the uploaded files

ChemDraw schemes are extracted from every uploaded file, as illustrated in figure 4-6a. These schemes are parsed by ChemScanner afterward. The output results of the parsing process, which contain interpreted reactions and molecules (figure 4-6b), are visualized by the "*Scanned Output*" component.

The "*Scanned Output*" component is comprised of two main functionalities: the results display and results edit.

4.3.1. Results display

The interpreted results from uploaded documents, which contain both reactions and molecules, are displayed as images to allow proof-reading the conversion of the files for correctness and completeness. Moreover, the ChemDraw JS[78] can be used as an option for the previewing of the original schemes from the uploaded files.

Practically, there are cases where the scheme does not have any reaction, or users intentionally want to retrieve only molecules from the uploaded schemes. Therefore, the "*Scanned Output*" interface provides two options for users to select: visualize only the molecules or visualize the reactions from the results.

Scan for molecules

Users can choose to only display molecules from the scanned results by selecting the "*Scan for molecules*" option in the "*Scan file*" dropdown from the "*Header Menu*". The interface, in this case, is shown in figure 4-7.

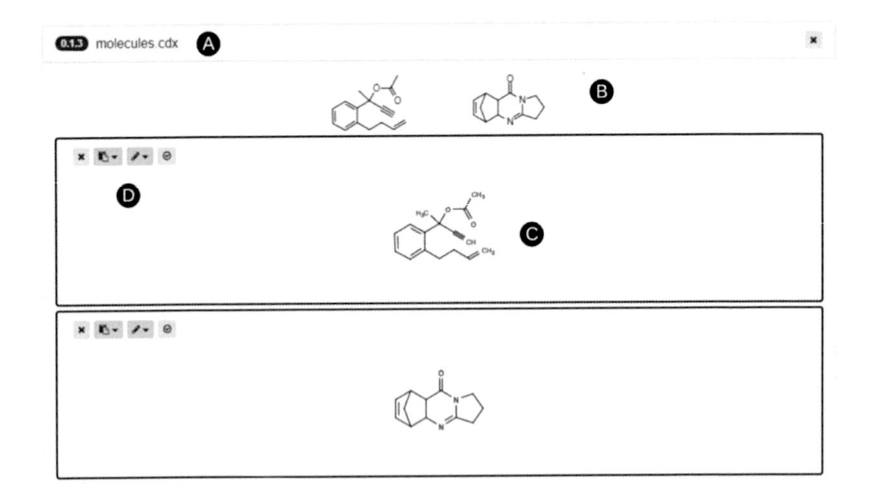

Figure 4-7 Displaying molecules from the output

The interface for displaying the scanned molecules can be divided into three parts

- **A**: the information of the uploaded file. In this example, the name of the uploaded file is "*molecules.cdx*", this was scanned by ChemScanner version "*0.1.3*".
- **B**: the image of the scheme from the uploaded file. This image is depicted from the ChemDraw JS[78]. In this case, the "*molecules.cdx*" file only contains one scheme.
- **C**: the visualized molecule from the results. ChemScanner detected two molecules from the "*molecules.cdx*" file, therefore two molecules are shown. The image of each molecule in this part is generated by OpenBabel[32].
- **D**: the menu of actions that are performed on each molecule. With the order from left to right

    - *Remove*: remove the molecule from the interface
    - *Copy to clipboard*: users can copy the SMILES or Molfile of the corresponding molecule.
    - *Edit Structure*: bring up the molecule editor to edit the scanned molecule if there is any mistake from `ChemScanner`. This feature is covered later in section 4.3.2.
    - *Select*: add the molecule into the selected list. The selected list is used for actions in the "*Header Menu*", as introduced in section 4.1

In addition to the features above, users can zoom and pan every image in the interface for better inspection. The zooming and panning feature is also applicable for the displaying of reactions; this feature is introduced in the next section.

Scan for reactions

Users can choose to display the reactions by selecting the "*Scan for reactions*" option in the "*Scan file*" dropdown from the "*Header Menu*". The interface, as described in figure 4-8, is similar to the displaying of the molecules with some difference below

- **A**: image of the interpreted reaction. The image is created by the backend of the interface by combining the image of each molecule in the reaction.
- **B**: the detected textual information of each reaction. The editing of the textual information that is displayed here is covered in section 4.3.2.

- Molecules of each molecule group (e.g., reactant group, product group) are distinguished by the number followed by the name of the molecule group. For example, "*Reactant 1*" indicates the first molecule, from the left, in the reactant group.
- In addition to the textual information of molecules, the text and the status (e.g., a successful reaction, failed reaction, planned reaction) of reaction are also displayed.

- **C**: the menu of actions that are performed on each reaction.
  - *Copy to clipboard*: copy the reaction SMILES, or the Molfile of each molecule from the reaction.
  - *Edit Structure*: bring up the molecule editor to edit the scanned molecule if there is any mistake from `ChemScanner`. The structural editing for the reaction is introduced in section 4.3.2.

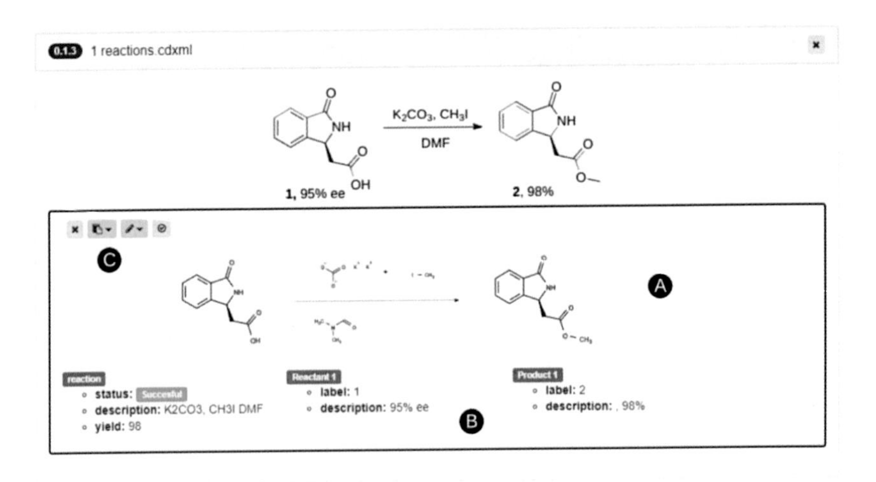

Figure 4-8 Displaying reactions from the output

4.3.2. Results edit

Add/remove reagents or solvents

The reagents and solvents of scanned reactions can be edited via the *Header Menu* interface. A library of more than 2000 most common reagents and catalysts were collected, and the contents can be added to a selected reaction.

The editing functionality is available when one reaction is added into the list of selected items. The following figure illustrates an example of adding a new reagent into the reaction from the scanned result in figure 4-8

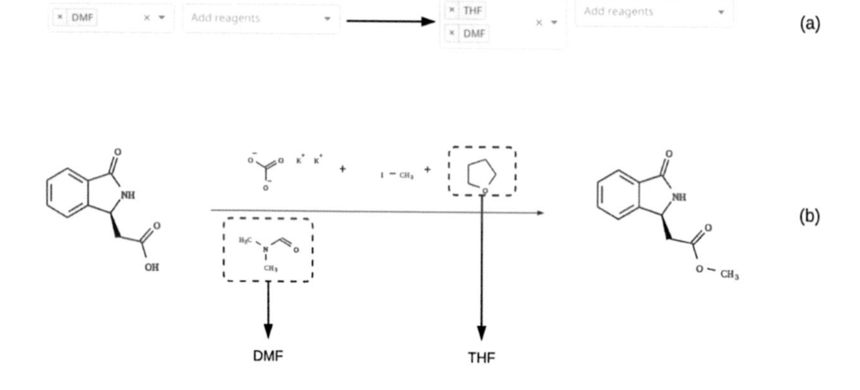

Figure 4-9 Illustration of reagent adding

When one reaction is selected, the "*Add solvents*" and "*Add reagents*" are enabled, as shown in figure 4-9a. The first image on the left indicates that there is one solvent, "*DMF*", is already used in the reaction. The "*X*" mark next to the entry can be used to remove the solvent from the reaction.

For illustration purposes, a new solvent "*THF*" is chosen from the solvent list. After the selection, the second image of figure 4-9a shows the current solvent list that is comprised of two solvents: "*THF*" and "*DMF*". Figure 4-9b describes the changing of the reaction after "*THF*" is added.

Text editing

Textual information of displayed results can be edited by double-clicking on the desired field, the editing process is illustrated in figure 4-10

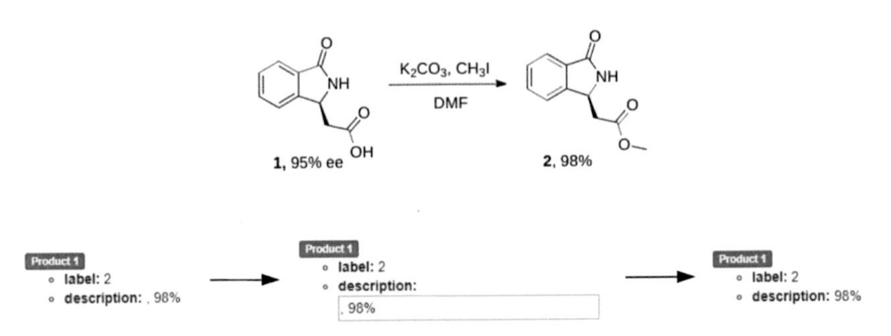

Figure 4-10 Illustration of text editing

In this example, the original text "`2, 98%`" is assigned the product of the reaction. ChemScanner detected the bolded text "**2**" is the label of the product, the bolded text "**2**" is assigned to the label field, and is removed from the original text string. The remaining text "`, 98%`" is recognized as the description of the molecule. The extra comma character is now redundant and can be removed by the editing of the *description* field, as shown in figure 4-10.

Structures editing

After the *text processing* of molecules from `ChemScanner`, as explained in chapter 3, the superatoms are replaced by corresponding structures. The coordinates of these structures are generated by the RDKit library[30]. Although the coordinate generation is based on the current coordinates of the molecule, the generation process is not perfect when processing complicated molecule, such as the molecule in figure 4-11b.

The *ChemScanner UI* provides the "*Clean up*" feature in the *Header Menu* to use the "*Clean up*" feature from ChemDraw. The "*Clean up*" feature is provided to the frontend by the ChemDraw JS[78]. When the "*Clean up*" feature is activated, a list of molecules is built like following

- If the selected list is empty, then all of the molecules, including molecules from reactions, are added to the list.
- Otherwise, every selected molecule and molecules from selected reactions are added to the list.

Using the list that is created above, the "*Clean up*" processing is applied to each molecule. For example, the molecule in figure 4-11c is the result of the "Clean up" processing of the molecule in figure 4-11b.

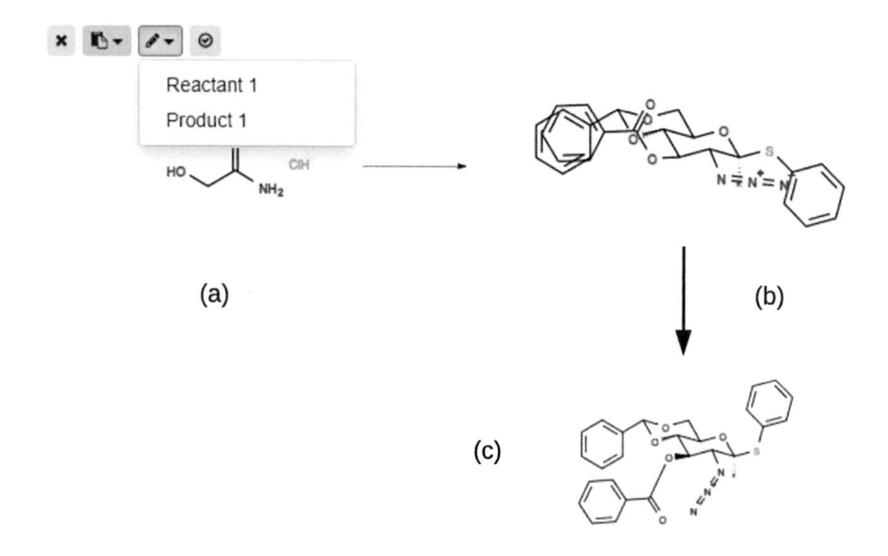

Figure 4-11 Illustration of structure editing

In addition to the "*Clean Up*" feature, users can manually edit the molecule structure via the action menu of each extracted item, as illustrated in figure 4-11a. In figure 4-11a, "`Reactant 1`" indicates the first molecule in the reactant group, and "`Product 1`" indicates the first molecule in the reactant group. When a molecule is chosen, a *Structure Editor* is shown up. The `ketcher-rails`[79] editor is employed as the primary molecule editor of the *ChemScanner UI*. Figure 4-12 illustrates the interface of the `ketcher-rails` editor when "`Product 1`" is chosen. The target molecule is updated after the editing if the user chooses to *Save* the change via the interface.

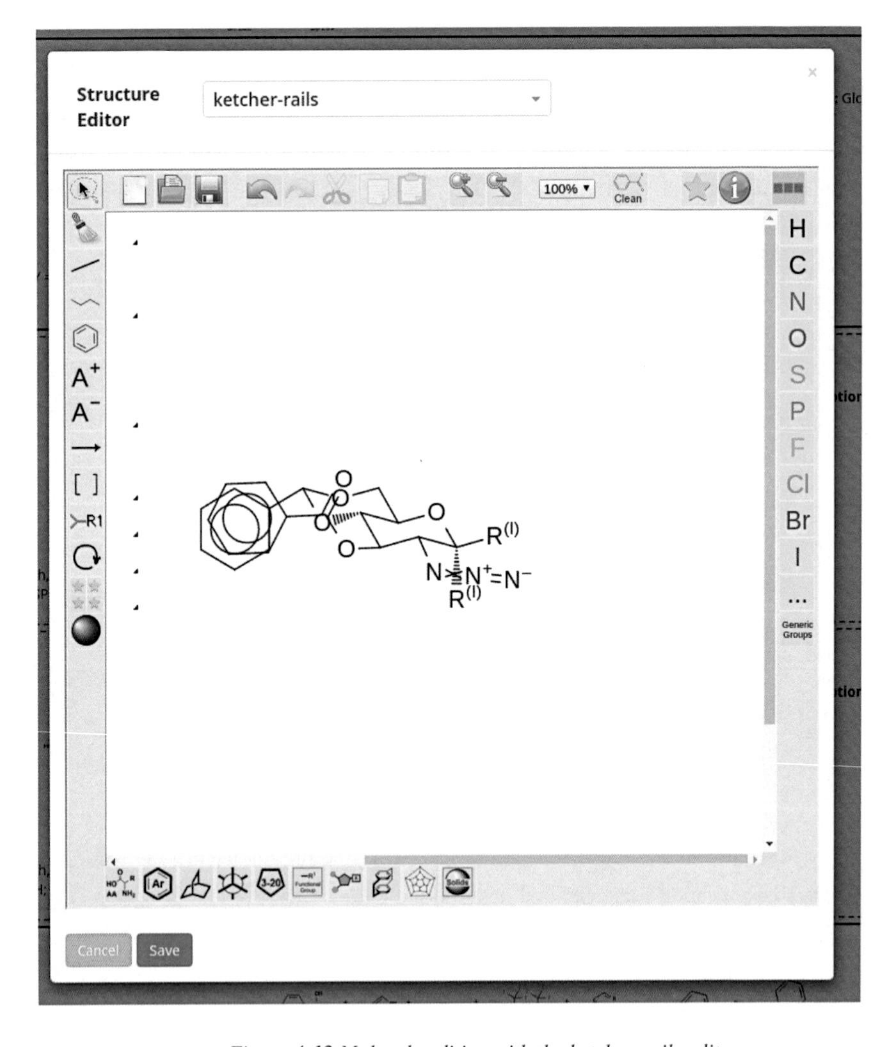

Figure 4-12 Molecule editing with the ketcher-rails editor

4.4 File Storage

4.4.1. Backend classes

The *File Storage* component is responsible for the management of uploaded files, including the schemes that are extracted from these uploaded files, as well as the results of these schemes after the *ChemScanner* processing.

ChemDraw schemes are extracted from the uploaded files; each uploaded file can contain multiple schemes, as shown in figure 4-13a. The uploaded files and the extracted schemes are saved into the database.

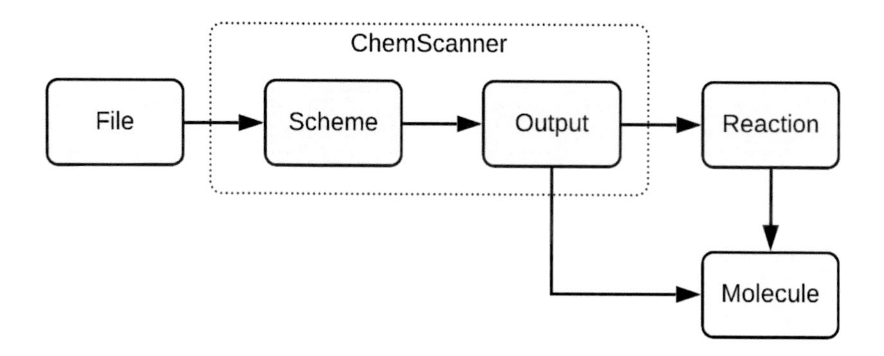

Figure 4-13 Processing flow in the backend

`ChemScanner` is used to process each extracted scheme to retrieve reactions and molecules from each scheme. The processing of `ChemScanner` can be applied to each scheme multiple times. The results of each processing, include detected reactions and molecules, are also saved into the database.

4.4.2. File Storage interface

The "*File Storage*" component is displayed when the "*File Storage*" option is chosen via the "*View*" dropdown button. The main interface, as illustrated in figure 4-14, is a tree grid consists of uploaded files as the first-level entries of the grid.

Schemes that are extracted from each file are displayed as children below each file (label **D** in figure 4-14). Each scheme has two children rows that display the summarize information of each scheme. The first row shows the number of molecules, while the second row shows the number of reactions within the scheme. If the file contains only one scheme, the scheme row is omitted.

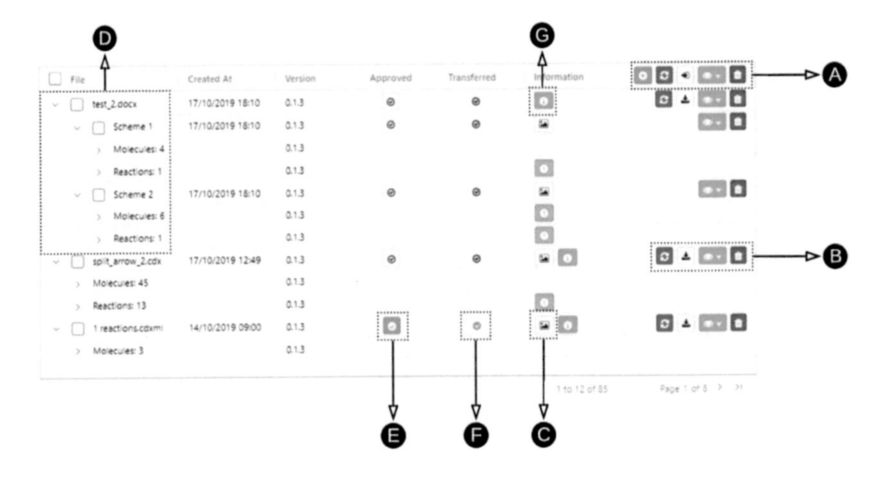

Figure 4-14 File Storage interface

Files and schemes can be selected using the checkbox on the same row. If a file is selected, all of its schemes are also selected. The grid action menu (label **A** of figure 4-14) contains actions that are applied to the selected entries within the grid. The grid action menu has five following actions that correspond to the icon from left to right

- **Upload files**: users can upload new files using this button. Unlike the "*Scan file*" feature, uploaded files using this action are uploaded only, without further processing by `ChemScanner`.
- **Rescan entries**: re-scan selected files or schemes. This action can be used with the "*upload files*" action to have the same effect as the "*scan files*" feature. This action is especially useful when the `ChemScanner` library is updated with new features and detection rules. The uploaded files can be scanned again with the new version of `ChemScanner`.

- **Import entries**: import selected entries into the Chemotion ELN[2]. If a file or scheme was already imported, a green icon is displayed as shown in label **F** in figure 4-14
- **Shows results**: display the result in the "*Scanned Outputs*" view. The action has two options
    - *Show molecules*: display only the molecules from the results of `ChemScanner`.
    - *Show reactions*: display reactions from the results of `ChemScanner`.
- **Delete entries**: delete selected files and schemes from the database.
    - If a scheme is deleted, its corresponding outputs are also deleted.
    - If a file is deleted, all of its schemes and outputs are also deleted.
    - If all the schemes of one file are deleted, this file still remains. A file is only deleted when users choose the delete the file.

Each selectable entry (file or scheme) in the grid also has its own action menu, with similar functionality (label **B** in figure 4-14).

- **Rescan entries**: Re-scan the corresponding file or scheme.
- **Download file**: download the corresponding file. This action is only available for file entry
- **Shows/hide results**: display or hide the result in the "*Scanned Outputs*" view. If the corresponding file or scheme is already displayed, this action hides its results from the "*Scanned Outputs*" view. It means that the results are removed from the view; the data still remain in the database. Similar to the action from the grid header, this action has two options: display the molecules or display the reactions.
- **Delete**: completely delete the file or scheme from the database. Its corresponding results are also deleted from the database.

Results approving

The approve button (label **E** in figure 4-4) is used to approve results from the `ChemScanner` processing of one file or scheme. If a file is approved, all of its schemes are also approved.

An approved scheme is considered that reactions and molecules from the `ChemScanner` outputs are appropriately interpreted. These approving data can be used for future enhancements of the `ChemScanner` library

Scheme previewing

The preview button (label **C** in figure 4-4) responsible for the displaying of the corresponding scheme without adding the scheme into the "*Scanned Outputs*" view. An example of previewing a CDXML file is illustrated in figure 4-15.

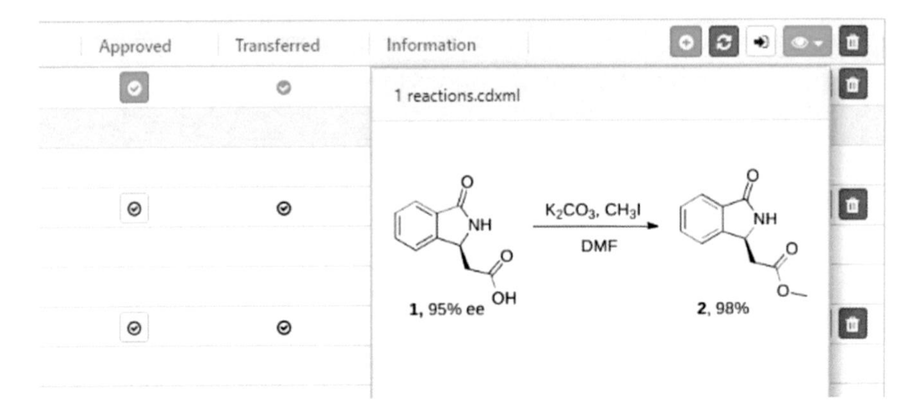

Figure 4-15 Scheme previewing

If the scheme belongs to a DOCX document, the image of the scheme is extracted directly from the DOCX contents, as introduced in chapter 2 and 3. Otherwise, an image generating process is initiated to depict the scheme's image using the ChemDraw JS. The image generating process is only activated one time per each scheme. The output image then is saved into the database so that it can be loaded if users when to use the *preview* feature again on the same scheme.

Metadata information

To support further management, the *ChemScanner UI* provides an interface for the viewing and editing of the metadata information, as shown in figure 4-15.

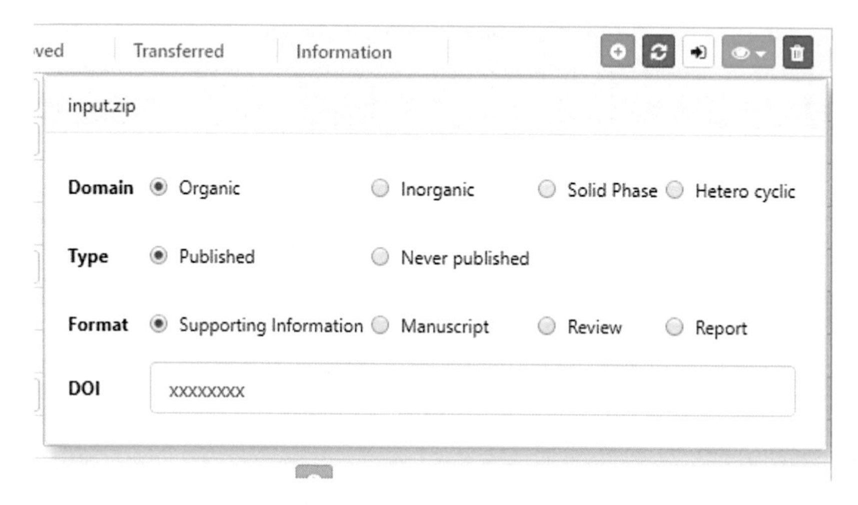

Figure 4-16 File additional information interface

The metadata information interface is shown using the "information" button (label **G** in figure 4-4). At the current state, four metadata attributes are supported by the *ChemScanner UI*, as shown in figure 4-15.

The metadata information is automatically parsed by the backend of the *ChemScanner UI* when an uploaded file is a compressed file, specifically a ZIP file. If the ZIP file contains a JSON[80] file, this JSON file is used to retrieve metadata attributes. The retrieved metadata information is shared between all other files that are nested within the ZIP container.

Users can also edit the parsed metadata information, or set the information for files that do not have the metadata information yet.

Import to the Chemotion Electronic Lab Notebook

When molecules and reactions from the results of ChemScanner are correctly retrieved and edited, they are ready for further processing for different purposes.

On the one hand, users can use the *Export* feature to export reactions and molecules into Excel or CML files. On the other hand, the *ChemScanner UI* provides an option for users to import reactions and molecules into the Chemotion Electronic Lab Notebook[2]. The import feature is available via the *import* action in the grid menu, as introduced in the previous section. Figure 4-16 shows the interface of reaction and molecule importing.

Figure 4-17 Import to Chemotion ELN

Users can choose to import reactions and molecules from selected files and schemes into an existed collection in the Chemotion ELN, or to create a new dedicated collection for the importing. Figure 4-18 describes the imported molecules and reactions after import the scheme in figure 4-8.

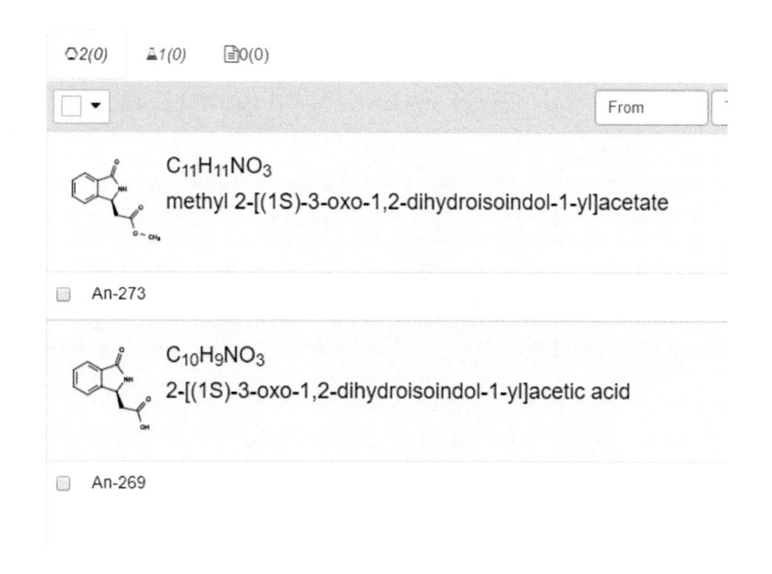

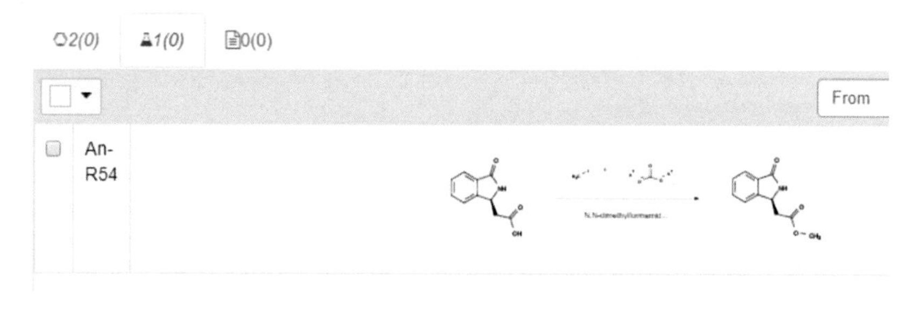

Figure 4-18 An example of importing to the Chemotion ELN

Chapter 5. Chemotion ELN integration

The *ChemScanner User Interface*, that is covered in chapter 4, is developed not only as a standalone web application but also being a part of the integration of `ChemScanner` into the Chemotion Electronic Lab Notebook (ELN). Besides the visualization of interpreted molecules and reactions from `ChemScanner`, the *ChemScanner User Interface* uses the `ketcher-rails`[79] editor that is provided by the ELN. Other than that, the *import* feature from the *ChemScanner UI* is directly connected with the database of the ELN, in order to import molecules and reactions from the results of `ChemScanner` into the database of the ELN.

This chapter introduces other parts of the integration process. First of all, an explanation of how molecules are searched in the database using the chemical structure search is given. Secondly, the interface that is used to display the computational properties of molecules is covered. Finally, the interface for green chemistry, which calculates and visualizes the green chemistry metrics[81] of each reaction, is introduced in section 5.3.

5.1 Chemical searching

Molecule searching is an essential feature for chemical databases. Therefore, a chemical structure search feature is developed in order to query a target molecule or fragment on a set of molecules, that are imported from the results of ChemScanner or other molecules from the Chemotion ELN. In general, the searching of molecular structures can be categorized into three types: Exact structure search, substructure search, similarity search.

5.1.1. Exact structure search

The exact structure is used to find the exact match of a whole molecule fragment from the query on a set of molecules in the database, despite the different visualization of the query molecule (e.g., the size of the molecule, the coordinates of atoms).

The exact structure searching is performed based on the unique molecule identifiers. For example, as introduced in chapter 2, canonical SMILES can be used as a molecule identifier to search for molecules within the same application database. However, different systems have different algorithms to generate the canonical SMILES of molecules, so that it cannot be used interchangeably between multiple chemical databases.

On the other hand, *InChI* and *InChIKey*, although they are not human-friendly formats, are widely-used as unique molecule identifiers because of their consistency across many chemical systems.

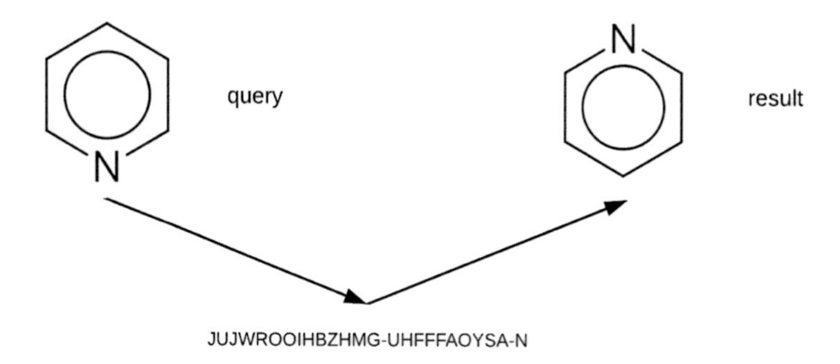

Figure 5-1 Illustration of exact structure search for pyridine

The *Chemotion ELN* database employs *InChiKey* as the molecule index for the exact structure search. Figure 5-1 illustrates an example of searching for *pyridine*; the *InChIKey* is generated from the query molecule. This *InChIKey* then is used to find the molecule that has the same *InChIKey* value. The resulting molecule is returned, although its structure is not exactly the same with the query molecule.

5.1.2. Substructure search

Substructure search is used for retrieving a set of all molecules that contain a given fragment, which is entered by the user as a substructure, including the exact input structure. The results from the substructure search are the superstructure of the query structure. The substructure search is useful for finding all available starting materials or products that contain a specific structure fragment.

A molecule can be represented as a simple undirected graph. It means that

- All the edges of the graph are bidirectional
- Two nodes cannot be connected by more than one edge.

Substructure matching is known as a subgraph isomorphism problem and is classified as an NP-Complete problem[82]. It means that the execution time for a substructure searching is, in the worst case, could not be expressed as a polynomial of the number of atoms and bonds. Therefore, in order to improve the speed of the searching, indexing the molecules in the database is a critical task for the performance.

The molecule in the database of the Chemotion ELN is indexed using the molecular fingerprint, as introduced in chapter 2. Whenever a molecule is created in the database, an FP2 fingerprint[38] is computed and saved into the database. The Chemotion ELN employs PostgreSQL[74] as the database of the web application, that supports up to 8-bytes for a column type. The FP2 fingerprint requires 1024-bit for each fingerprint entry; therefore, each fingerprint is split into sixteen 8-bytes, or 64-bits, of integers. Figure 5-2 illustrates how the fingerprint is saved into the PostgreSQL database.

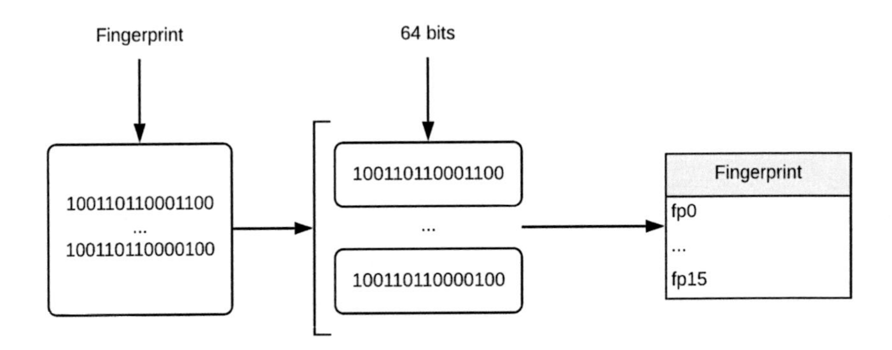

Figure 5-2 Saving fingerprint into the PostgreSQL table

By adapting molecular fingerprint as for the indexing of molecules in the database, an important characteristic of the molecular fingerprint can be employed: If one fragment is a substructure of a molecule, every bit that is set in the fingerprint of the fragment also is set in the fingerprint of the molecule. Moreover, since bitwise operations are considered one of the fastest operations in modern processors, the usage of the fingerprint comparison can significantly decrease the execution time for substructure searching.

Due to the nature of the OpenBabel FP2 fingerprint, the bit position that is calculated by the hash function of the FP2 fingerprint, it not always unique. There are cases that multiple fragments are set in the same bit position. So that the results returned by the fingerprint comparison can have false positives candidates, it means that there are molecules in the results that do not match the substructure search. Therefore, the Chemotion ELN employs the fingerprint in the screening phase for the substructure search, as described in figure 5-3.

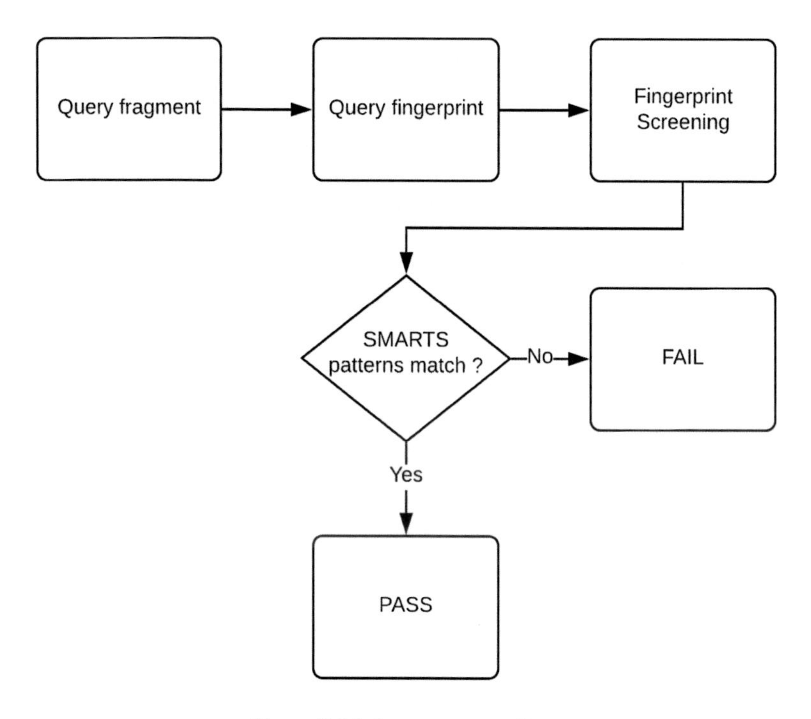

Figure 5-3 Substructure searching process

When a user creates a fragment as the query for the substructure search, an FP2 fingerprint is generated from the input fragment. This fingerprint then is applied for the fingerprint screening phase; the query fingerprint is compared with the molecular fingerprints from the database. The output of the screening process is used by the substructure matcher. The Chemotion ELN uses the SMARTS substructure matcher that is provided by the OpenBabel library.

Figure 5-4 illustrates an example of the substructure searching of *phenol* on one collection from the Chemotion ELN.

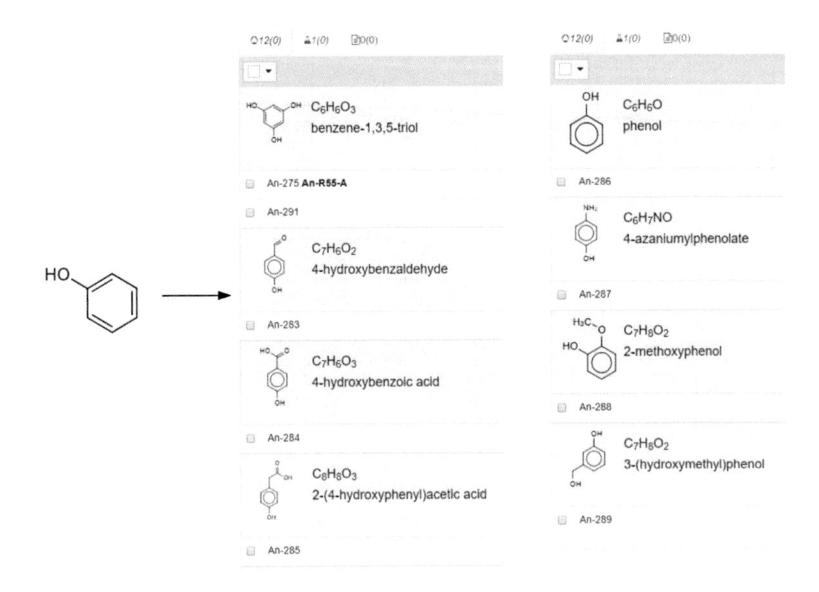

Figure 5-4 The results from the substructure search of Phenol

5.1.3. Similarity search

Molecules that have similar structures tend to have similar properties[83] (e.g., biological activity, solubility). Based on this principle, the molecular similarity can be used in predicting the molecule properties or drug design studies by searching on the database for known active compounds that are similar to the target molecule. In order to serve the need for molecular similarity, the Chemotion ELN provides the similarity search on its molecule databases.

To achieve high performance in similarity search, the Chemotion ELN employs the OpenBabel FP2 fingerprint that has already existed in the database to support the substructure search. The most well-known similarity metric for comparing chemical structures using the molecular fingerprints is the *Tanimoto coefficient* [84].

Assuming two fingerprints of two molecules A and B, with A is the fingerprint of the query molecule, and B is the fingerprint of a molecule in the database that is compared with the query molecule. Three values are defined as follows

- N_A: the total number of bits that are set in the fingerprint A.
- N_B: the total number of bits that are set in the fingerprint B.
- N_{AB}: the total number of bits that are set in both fingerprints A and B.

With these notations, the *Tanimoto coefficient T* is calculated by the formula:

$$T = \frac{N_{AB}}{N_A + N_B - N_{AB}}$$

Based on the formula, the value of the *Tanimoto coefficient T* has the range from 0 to 1, or from no similarity to perfect similarity. For path-based fingerprints, the FP2 fingerprint specifically, if the *Tanimoto coefficient* satisfies the condition $T > 0.85$, the two structures are usually considered as similar, and these two structures have similar activities[85].

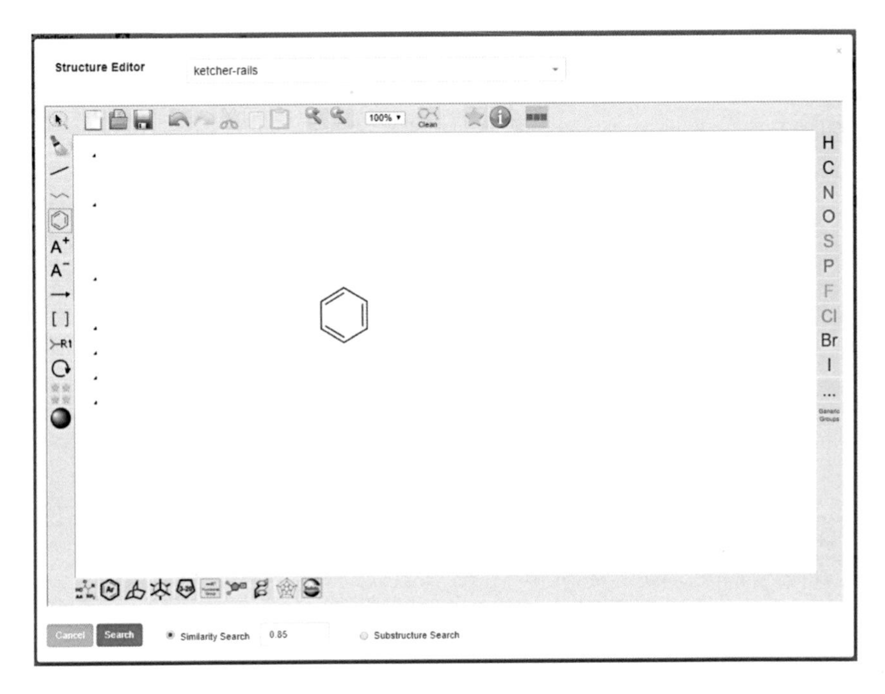

Figure 5-5 Structure search interface

Figure 5-5 shows the interface of the similarity search and substructure search from the Chemotion ELN. Users can draw the query structure using the `ketcher_rails` editor[79], and choose the type of structure searching from the interface. Although two structures are considered similar if $T > 0.85$, users can still change to the threshold of the *Tanimoto coefficient*, as shown in figure 5-4.

In order to improve the searching time, whenever a molecule is created in the database, an FP2 fingerprint[38] is generated and saved to the database. The total number of bits that are set in the generated fingerprint is also calculated, and saved into the `num_set_bits` field in the `Fingerprint` table from the database. When a user activates the similarity search, the Chemotion ELN follows the procedure, which is described in figure 5-6, to search for molecules that are similar to the query molecule.

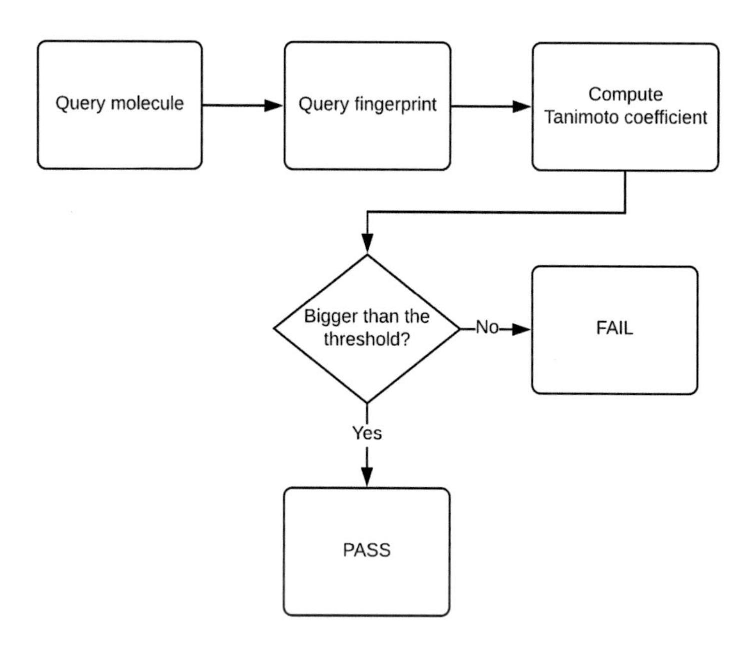

Figure 5-6 Similarity searching process

Identical to the substructure search, an FP2 fingerprint is generated from the input molecule. The total number of bits that are set in this fingerprint, N_A, is calculated. Then with each fingerprint B in the database, the corresponding N_B and N_{AB} are computed. Also, the *Tanimoto coefficient* is derived from N_A, N_B, and N_{AB}. If the calculated *Tanimoto coefficient* is bigger than the input threshold, the corresponding molecule is considered as a similar molecule with the query molecule on the input threshold.

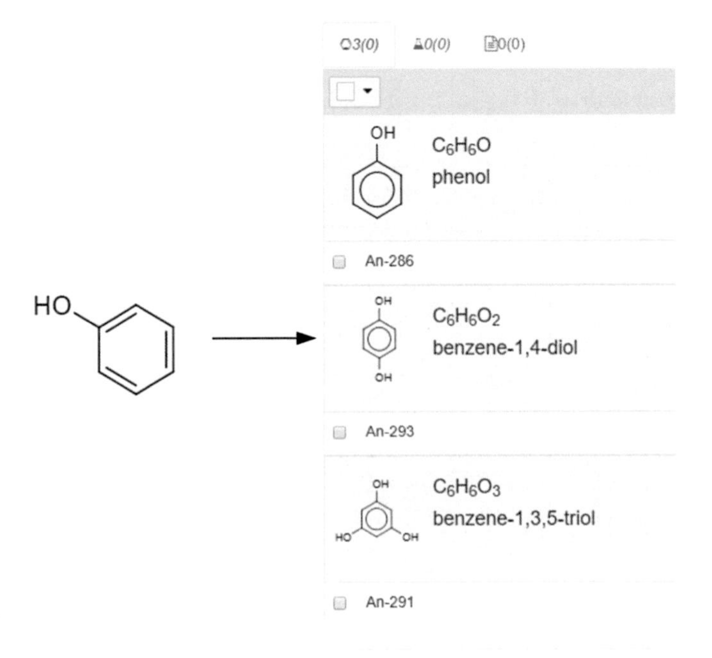

Figure 5-7 The results from the similarity search of Phenol

Figure 5-7 illustrates an example of the similarity searching of *phenol* on the same collection in figure 5-4. The threshold of the *Tanimoto coefficient* in figure 5-7 is set to `0.85`.

5.2 Computational molecule properties interface

Organic semiconducting materials are used in Organic Light-Emitting Diodes[86] (OLED), also known as organic electroluminescent (organic EL), have therefore become an essential material for modern electronic displays due to its advantages[87] (e.g. lightweight, better color quality, higher scan rate). It is known that the energy efficiency is improved by increasing the electron mobility of the OLED[88,89]. Using this principle, *Friedrerich et al.*[90] employed a multiscale simulation approach to find ways to systematically improve the charge carrier mobility of materials for the Electron Transport Layers (ETLs). During the simulation progress, not only the electron mobility of the molecule is calculated, but also intermediate properties are generated. The *computational molecule properties interface* integrates the simulation process into the Chemotion ELN to improve the reusability of these results.

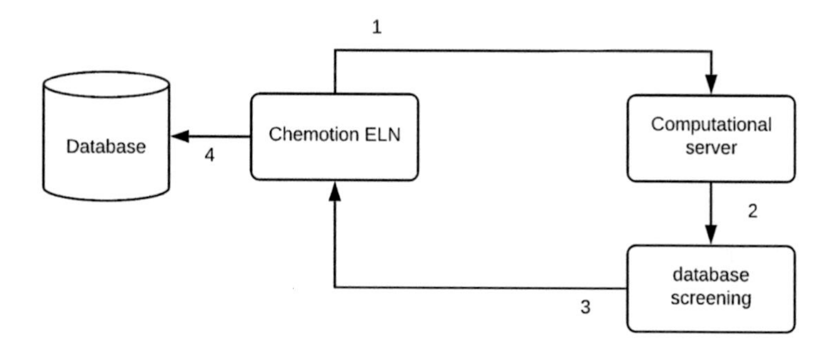

Figure 5-8 Data flow for computing molecule properties

The *computational molecule properties interface* provides an API to save the output of the computation into the database, and to visualize the results from one or more molecules. Figure 5-8 describes how molecules are computed. Firstly, a molecule is drawn in the Chemotion ELN using the `ketcher_rails` editor, a molecule SMILES is generated and is sent to the *computational server* module.

The *computational server* is an API server that is built by Python[91] using the Flask framework[92]. After receiving the request from the Chemotion ELN, the server activates the *database screening* module to simulate the received molecular SMILES. The workflow of the *database screening* module is described in figure 5-9.

Firstly, the *database screening module* uses the *molconvert*[93] to generate the 3D coordinates (XYZ file) from the input SMILES, received from the *computational server* module. The MOPAC software[94] then uses the PM6 method for the semi-empirical optimization from the generated 3D coordinates. The optimization from MOPAC is needed for a faster optimization with Turbomole[95] in the next step.

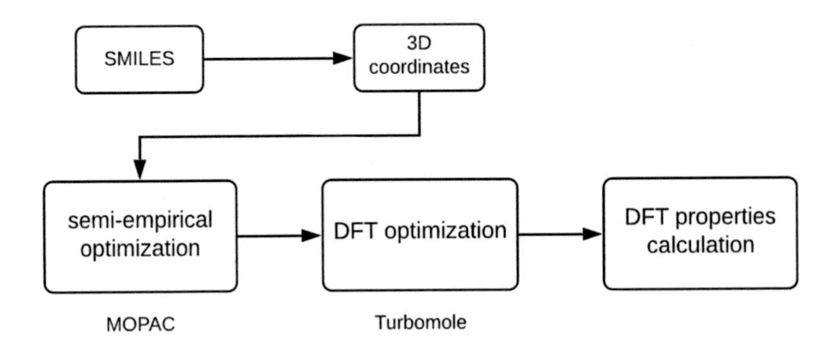

Figure 5-9 The workflow of database screening module

Turbomole takes the pre-optimized geometry of the molecule from MOPAC and performs the DFT optimization in neutral state as well as single-point calculations in three different charge states: neutral, positively charged and negatively charged. On each level, the properties of the molecule are calculated using the B3-LYP exchange-correlation functional and the def2-SV(P) basis set.

The results from these three calculations are combined to calculate the following properties of the molecule: the maximum electrostatic potential, the minimum electrostatic potential, the mean electrostatic potential, the mean-absolute electrostatic potential on a surface of the molecule. These electrostatic potential properties are calculated in millivolt (mV).

In addition, the highest occupied molecular orbital (HOMO), lowest unoccupied molecular orbital (LUMO), ionization potentials (IP), and electron affinities (EA) are derived in electronvolt (eV). The dipole moment, in Debye, is also measured.

When the simulation is finished, the results of the *database screening module* are sent back to the Chemotion ELN and are saved into the database, as illustrated at step 3 of figure 5-8. The output is stored in the database and can be used for the visualization in the interface of the Chemotion ELN, as illustrated in figure 5-10.

MaP	MiP	MeP	MeAbsP	HOMO	LUMO	IP	EA	Dipol	Date
398	-436	-10	94	-6.07	-0.18	8.29	-1.96	1.38	9 Jul 2018

Figure 5-11 Summary of computed properties of phenol

The *compute* button is used to activate the calculation process, as described in figure 5-8. For illustration purposes, figure 5-10 shows a summary of the calculation result from the simulation of *phenol*. The *MaP*, *MiP*, *MeP*, *MeAbsP* are the abbreviations for the maximum, minimum, mean, and mean-absolute electrostatic potential, respectively. These computational results can be visualized on the graph, which can be customized through the template configuration, as shown in figure 5-11.

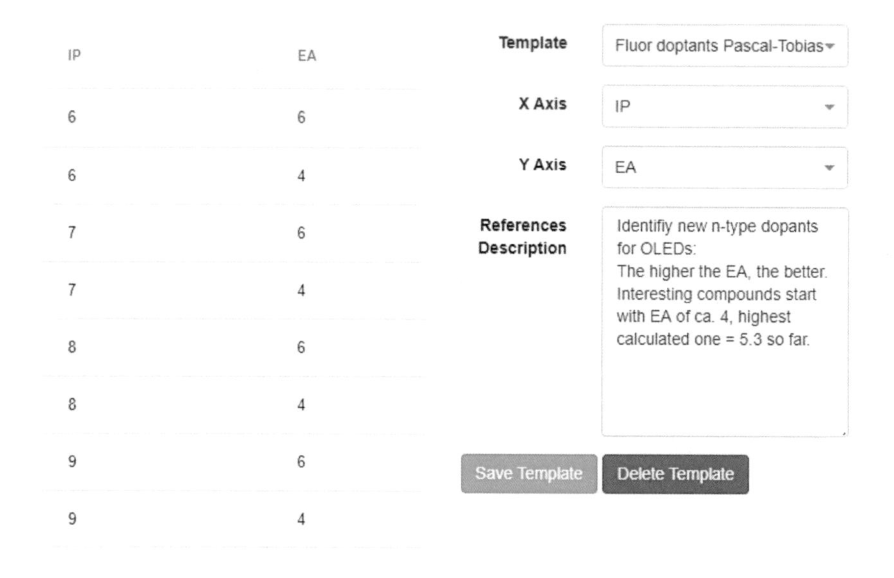

IP	EA
6	6
6	4
7	6
7	4
8	6
8	4
9	6
9	4

Figure 5-10 Template for visualization graph

The template configuration interface has a predefined default template that could be used to visualize the graph for materials of the Electron Transport Layer. Users can customize the graph by changing the "*X-Axis*" and "*Y-Axis*" parameters. The interface provides six options for each parameter: HOMO, LUMO, IP, EA, Dipole, and electrostatic potential (ESP).

Moreover, users can also define a list of *reference points*, as shown on the left side of figure 5-10. The figure shows a list of eight *reference points* that are displayed in the graph together with the point that represents the properties of the molecule.

Finally, with the data of the molecule, a graph is shown with the corresponding template configuration. Using the Chemotion ELN, users can not only visualize data from one molecule but also with multiple molecules. Figure 5-12 illustrates the visualization of the data of 21 molecules with 8 references point. The graph is generated by using the settings from the template of figure 5-11.

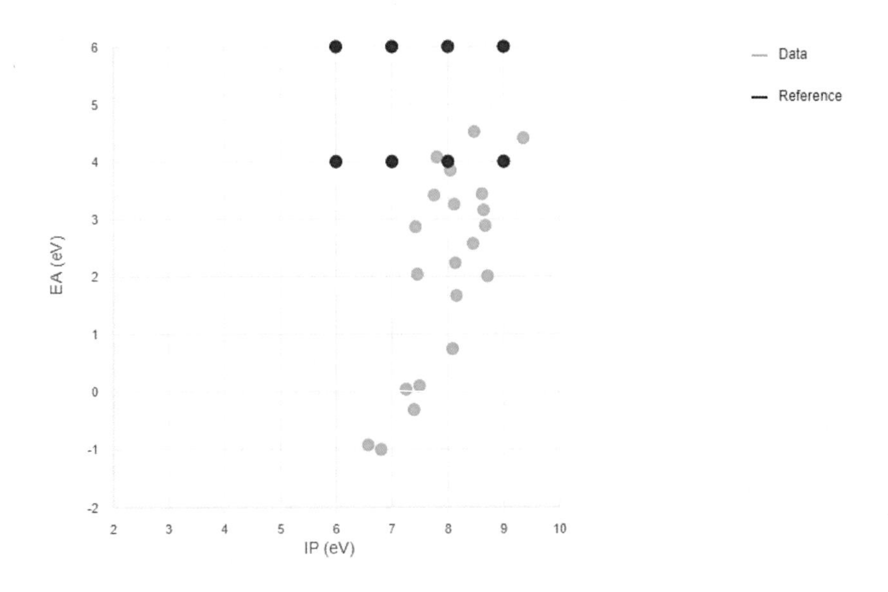

Figure 5-12 Visualization from template and data

5.3 Green Chemistry

Green Chemistry[96], or sustainable chemistry, is widely known as a growing area of chemistry focusing on the designing of chemical products that are more environmentally-friendly for the chemical industry. The green chemistry can be used as a guide for the chemical synthesis process for minimizing the waste and the generation of hazardous substances.

The inventor of the term "green chemistry" published a set of twelve design principles[96], as the practice guidance for the green chemistry

1. **Prevention**: Preventing waste is better than waste treatment or remediation.
2. **Atom Economy**: Attempt to maximize the incorporation of all materials used in the synthesis into the product, so that less waste is generated. This metric is covered in detail in the next section
3. **Less Hazardous Chemical Syntheses**: Synthetic methods should be designed to avoid generating substances that are toxic to humans and the environment.
4. **Designing Safer Chemicals**: The products of the synthesis process should be as non-toxic as possible.
5. **Safer Solvents and Auxiliaries**: Avoid the use of auxiliary substances whenever possible. If they must be used, they should be less hazardous as possible.
6. **Design for Energy Efficiency**: Synthetic methods should be designed to use as little energy as possible.
7. **Use of Renewable Feedstocks**: Renewable materials are preferred that non-renewable materials, if applicable.
8. **Reduce Derivatives**: More synthesis steps can generate more waste and require additional reagents so that shorter syntheses should have higher priority.
9. **Catalysis**: prefer catalytic reagents than stoichiometric reagents.
10. **Design for Degradation**: Chemical products should be designed to be degradable so that they do not pollute the environment.
11. **Real-time analysis for Pollution Prevention**: Development of analytical methodologies for real-time monitoring and controlling to prevent further pollution.
12. **Inherently Safer Chemistry for Accident Prevention**: Substances that are used in the process should be selected to have low-risk potential.

In order to quantify the greenness of chemistry processes, various green chemistry metrics are developed and proposed. Two earliest, useful, and most well-known metrics among them are the atom economy and the environmental (E) factor metrics.

5.3.1. Atom Economy

Being the second principle of green chemistry, the atom economy, which was developed by Barry Trost[97], describes what atoms of the reactants remain in the final products. The atom economy P_{ae} of a reaction is calculated by dividing the molecular weight of the reaction products by the sum of the molecular weights of all reactants in the reaction, as illustrated by the formula

$$P_{ae} = 100\% \times \frac{M_{product}}{\sum M_{reactant}}$$

With $M_{product}$ is the molecular weight of the product and $M_{reactant}$ is the molecular weight of each reactant.

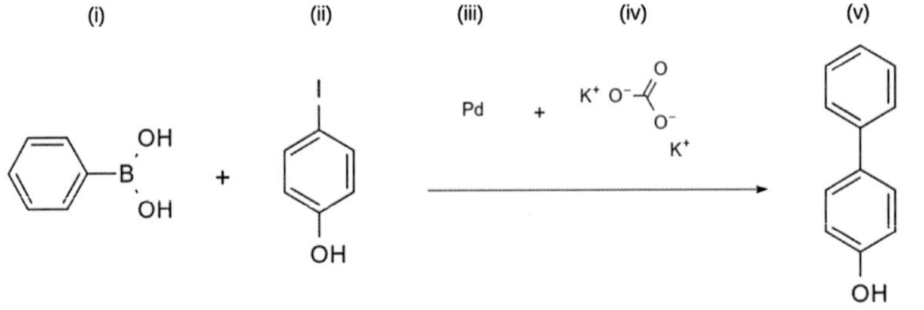

Figure 5-13 Example reaction

For example, consider the reaction[98] in figure 5-13, the molecular weight of each molecule in the reaction is listed below

- (i) : M_{i} = 121.9 g/mol
- (ii) : M_{ii} = 220.0 g/mol
- (iii) : M_{iii} = 106.4 g/mol
- (iv) : M_{iv} = 138.2 g/mol
- (v) : M_{v} = 170.2 g/mol

The atom economy of the reaction is

$$P_{ae} = 100\% \times \frac{170.2}{121.9 + 220 + 106.4 + 138.2} = 29\%$$

5.3.2. Environmental (E) factor

The environmental (E) factor[99] was introduced by Roger A. Sheldon in 1992; the metric is one of the most flexible and well-known in green chemistry.

The E-factor is obtained by dividing the mass of waste by the mass of product

$$E = \frac{M_{waste}}{M_{product}}$$

M_{waste} is the mass of the waste and $M_{product}$ is the mass of the product. Waste is defined as every compound that incorporated into the reaction except the desired product and water. Figure 5-14 describes the *E-factor* values in the chemical industry, calculated by Sheldon.

Industry Segment	E factor (kg waste per kg product)
Oil refining	< 0.1
Bulk chemicals	< 1 - 5
Fine chemicals	5 - 50
Pharmaceuticals	25 - > 100

Figure 5-14 E-factors in the chemical industry

Recently, *Roschangar et al.*[100] suggested two variations of the *E-factors*, that can be used depends on the stage of the development of the synthesis process: the *simple E-factors* (sEF) and the *complete E-factors* (cEF)

- The *simple E-factor* excludes the mass of solvents and water from the calculation formula

$$sEF = \frac{M_{materials} + M_{reagents} - M_{product}}{M_{product}}$$

- While the *complete E-factor* take into account of all substances in the reaction

$$cEF = \frac{M_{materials} + M_{reagents} + M_{solvents} - M_{product}}{M_{product}}$$

5.3.3. Green Chemistry Interface

The Green Chemistry interface provides an interface to display the two green chemistry metrics that are introduced above: the *atom economy* and the *E-factor*, as illustrated in the figure below

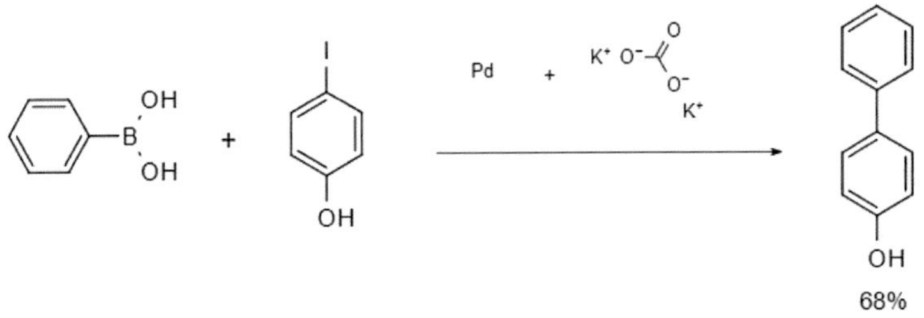

Pd + K+ O- O O- K+

68%

Scheme | Properties | References | Analyses | Green Chemistry

Simple E factor (sEF)	Complete E factor (cEF)	Custom E factor	Atom economy (AE)	Custom Atom economy
371.76	371.76	371.76	0.29	0.29

Figure 5-15 Green Chemistry Interface

Figure 5-15 shows the Green Chemistry interface of the same reaction in figure 5-10; the *atom economy* is calculated and is equal to 0.29, which is the same value that was already calculated in section 5.3.2.

The *simple E-factor* and the *complete E-factor* are also calculated and shown in the green chemistry interface. In figure 5-15, the value of the *simple E-factor* is equal to the *complete E-factor* because of the missing of solvent in the illustrated reaction.

In addition, the Green Chemistry interface provides two options: the *custom E-factor* and the *custom atom economy*. The calculation of these two metrics can also be customized via the interface, as shown in figure 5-16.

Starting Materials	Mass	Volume	Moles	Equiv.	Recyclable	Coeff
An-329 phenylboronic acid	0.1220	0.0000	0.0010	1.0000	☐	1
An-330 4-iodophenol	0.2200	0.0000	0.0010	0.9994	☐	1

Reactants	Mass	Volume	Moles	Equiv.	Recyclable	Coeff
palladium	1.0642	0.0000	0.0100	9.9942	☐	1
dipotassium;carbonate	41.4616	0.0000	0.3000	299.8268	☐	1

Products	Mass	Volume	Moles	Equiv.	Waste	Coeff
An-331 **An-R64-A** 4-phenylphenol	0.1150	0.0000	0.0007	0.6753	☐	1

Figure 5-16 Customization of the Green Chemistry metrics

The *custom E-factor* takes into account molecules that are *recyclable* and products that are considered as *waste*

$$cuEF = \frac{M_{materials} + M_{reagents} + M_{solvents} - M_{product} - M_{waste} - M_{recyclable}}{M_{product}}$$

The *custom atom economy* P_{cae} is calculated by taking into account the *coefficient*, which is shown as the "*Coeff*" field in figure 5-16, of each compound. The molecular weight of each molecule is multiplied with the *coefficient*

$$P_{cae} = 100\% \times \frac{M_{product} \times C_{product}}{\sum(M_{reactant} \times C_{reactant})}$$

Chapter 6. Summary and conclusions

The extracting of chemical knowledge is a complicated challenge, especially with the vast majority of data from publications, patents, and databases. The research project contributes to the development of chemical databases by presenting a computer-aided system for chemical data retrieving.

The ChemScanner library is used to interpret ChemDraw-related file formats. The library is a complementary to current state-of-the-art OCSR approaches. Most of the molecular recognition challenges can be handled appropriately by using the library. The ability to extract embedded ChemDraw schemes from the DOC and DOCX files facilitates users in building chemical databases from accessible documents.

The ChemScanner library is developed to be able to recognize and detect reactions based on textual information. The generation of new reactions and molecules provides more options for researchers and organizations to create more reaction databases. The reaction databases are useful not only for synthetic chemists to design synthesis routes but also applicable for developing new machine learning applications in retrosynthesis and reaction prediction.

The ChemScanner User Interface provides a reviewing and editing interface for human intervention. Together with the management interface, users have a feasible way to review the output of the current ChemScanner library, to inspect the results from the previous version of ChemScanner, and to allow incorrect conversion can be applied again later in the new releases of the ChemScanner library.

Molecule and reaction databases can be created directly via the importing into the Chemotion ELN database via the ELN-integration interface. Molecules and reactions in the database then can be quickly retrieved via the structure search, including substructure search and similarity search. Moreover, the computational properties of one molecule can be calculated via the computational molecule interface, or the greenness of one reaction can be simulated and displayed in the green chemistry interface. These additional properties are stored within the ELN and can be used later to accelerate further research and development.

Further Impact

The ChemScanner library may also contribute to the retrieval of reaction and their derivatives from the textual information. Assuming an Optical Chemical Structure Recognition (OCSR) software that is perfectly built for molecule recognition, text recognition, and graphic recognition. In order to extract embedded reactions or generate new molecules and reactions based on graphic patterns and textual information, the OCSR software can also use the rules from the open-source ChemScanner library.

Publishers of organic chemistry journals may also be interested in the proposed solutions from the research project. In fact, the research project has been recognized by the Beilstein journal[101]. The journal has initiated a collaboration to employ the ChemScanner system as part of their data retrieving system.

Areas for future work

Although the research project tries to cover various aspects of the chemical data retrieving process, there are still opportunities for enhancements.

There are more complicated schemes that the ChemScanner library needs to be able to interpret correctly. Figure 6-1 illustrates a generalized reaction that can be interpreted into eleven reactions based on the values of the Rgroup. To resolve this example and other similar scenarios, the ChemScanner library must have the ability to detect the Rgroup while comparing two structures. For example, by comparing the general product, the 3ba-la molecule, the corresponding Rgroup with the product 3ba is the Methyl group (Me). Each product would have a corresponding starting material and yield.

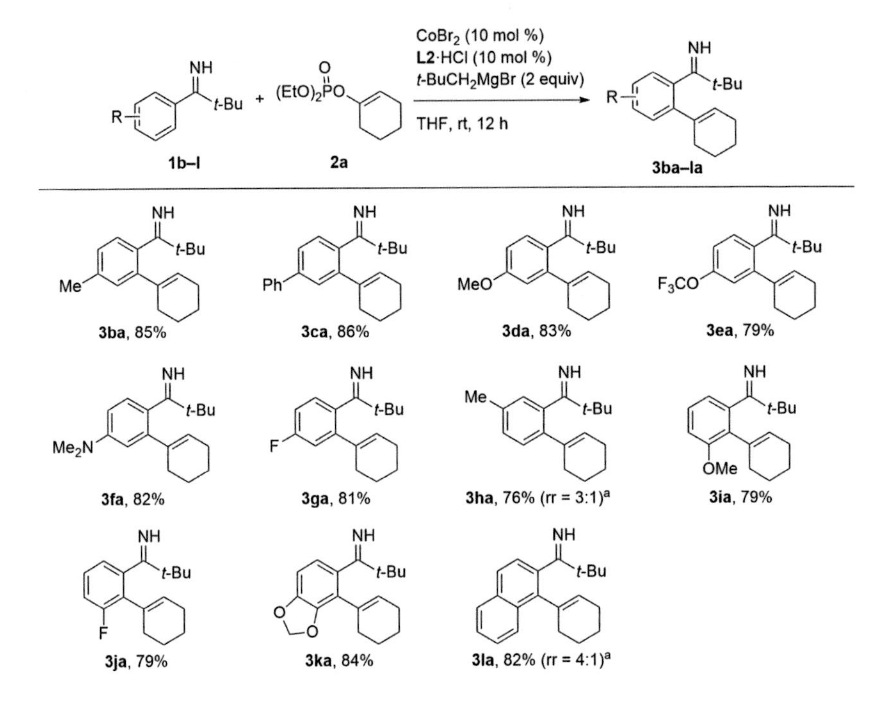

Figure 6-1 Example of a complicated ChemDraw scheme

On the other hand, the ChemScanner User Interface can have a confidence rating on each result from the ChemScanner library. This rating would help decrease the reviewing time significantly for users. Other improvements can be listed in the management interface such as the sorting the filtering data based on the metadata information, or adding new green metrics for the green chemistry interface (e.g., effective mass yield, carbon efficiency, reaction mass efficiency, the EcoScale[102])

Chapter 7. Abbreviations

%	Percent
°C	Temperature in Celsius
1D	One Dimensional
2D	Two Dimensional
3D	Three Dimensional
AE	Atom Economy
A.I.	Artificial Intelligence
API	Application Programming Interface
CDX	ChemDraw Binary
CDXML	ChemDraw XML
CER	Chemical Entity Recognition
CFB	Compound File Binary
CML	Chemical Markup Language
CNER	Chemical Named Entity Recognition
CRF	Conditional Random Field
CTAB	Connection Table
DT	Decision Tree
EA	Electron Affinity
ELN	Electronic lab notebook
ETL	Electron Transport Layer
ESP	Electrostatic Potential
FIB	File Information Block
HOMO	Highest Occupied Molecular Orbital

HTML	Hypertext Markup Language
HTTP	Hypertext Transfer Protocol
InChI	IUPAC International Chemical Identifier
IP	Ionization Potential
IUPAC	International Union of Atomic and Molecular Physical Data
JSON	JavaScript Object Notation
KIT	Karlsruhe Institute of Technology
LUMO	Lowest Unoccupied Molecular Orbital
MEMM	Maximum-entropy Markov Model
ML	Machine Learning
NLP	Natural Language Processing
OCR	Optical Character Recognition
OCSR	Optical Chemical Structure Recognition
OLE	Object Linking and Embedding
OOXML	Office Open XML
OSR	Optical Structure Recognition
PDF	Portable Document Format
POS	Part-of-speech
RF	Random forest
SMARTS	SMILES Arbitrary Target Specification
SMILES	Simplified Molecular Input Line Entry Specification
SPA	Single-Page Application
SVG	Scalable Vector Graphics
SWIG	Simplified Wrapper and Interface Generator

UI	User Interface
URL	Uniform Resource Locator
USPTO	United States Patent and Trademark Office
UV	Ultraviolet
WWW	World Wide Web
XML	Extensible Markup Language

Chapter 8. Literatures

1. Li Z, Wan H, Shi Y, Ouyang P. Personal Experience with Four Kinds of Chemical Structure Drawing Software: Review on ChemDraw, ChemWindow, ISIS/Draw, and ChemSketch. J Chem Inf Comput Sci. 2004 Sep 1;44(5):1886–90.

2. Tremouilhac P, Nguyen A, Huang Y-C, Kotov S, Lütjohann DS, Hübsch F, et al. Chemotion ELN: an Open Source electronic lab notebook for chemists in academia. J Cheminformatics. 2017 Sep 25;9(1):54.

3. Krallinger M, Rabal O, Lourenço A, Oyarzabal J, Valencia A. Information Retrieval and Text Mining Technologies for Chemistry. Chem Rev. 2017 Jun 28;117(12):7673–761.

4. Eltyeb S, Salim N. Chemical named entities recognition: a review on approaches and applications. J Cheminformatics. 2014 Apr 28;6:17.

5. Martin E, Monge A, Duret J-A, Gualandi F, Peitsch MC, Pospisil P. Building an R&D chemical registration system. J Cheminformatics. 2012 May 31;4(1):11.

6. PubChem. PubChem [Internet]. [cited 2019 Oct 26]. Available from: https://pubchem.ncbi.nlm.nih.gov/

7. ChemSpider [Internet]. [cited 2019 Oct 26]. Available from: https://www.chemspider.com/

8. SciFinder [Internet]. [cited 2019 Oct 26]. Available from: https://scifinder.cas.org/

9. Grego T, Pesquita C, Bastos HP, Couto FM. Chemical Entity Recognition and Resolution to ChEBI [Internet]. International Scholarly Research Notices. 2012 [cited 2019 Oct 26]. Available from: https://www.hindawi.com/journals/isrn/2012/619427/

10. Rocktäschel T, Weidlich M, Leser U. ChemSpot: a hybrid system for chemical named entity recognition. Bioinforma Oxf Engl. 2012 Jun 15;28(12):1633–40.

11. Lowe DM, Corbett PT, Murray-Rust P, Glen RC. Chemical name to structure: OPSIN, an open source solution. J Chem Inf Model. 2011 Mar 28;51(3):739–53.

12. Park J, Rosania GR, Shedden KA, Nguyen M, Lyu N, Saitou K. Automated extraction of chemical structure information from digital raster images. Chem Cent J. 2009 Feb 5;3:4.

13. Algorri M-E, Zimmermann M, Friedrich CM, Akle S, Hofmann-Apitius M. Reconstruction of Chemical Molecules from Images. In: 2007 29th Annual International Conference of the IEEE Engineering in Medicine and Biology Society. 2007. p. 4609–12.

14. Filippov IV, Nicklaus MC. Optical Structure Recognition Software To Recover Chemical Information: OSRA, An Open Source Solution. J Chem Inf Model. 2009 Mar 23;49(3):740–3.

15. Zimmermann M. Chemical Structure Reconstruction with chemoCR. In 2011.

16. USPTO Bulk Downloads: Patent Grant Full Text with Embedded Images [Internet]. Available from: https://www.google.com/googlebooks/uspto-patents-redbook.html

17. CLiDE [Internet]. [cited 2019 Oct 22]. Available from: http://www.keymodule.co.uk/products/clide/index.html

18. ChemoCR [Internet]. Fraunhofer Institute for Algorithms and Scientific Computing SCAI. [cited 2019 Oct 22]. Available from: https://www.scai.fraunhofer.de/en/business-research-areas/bioinformatics/products/chemocr.html

19. Imago OCR [Internet]. [cited 2019 Oct 22]. Available from: https://lifescience.opensource.epam.com/imago/index.html

20. Pearson N. Enhancing the User Experience for Wiley Chemistry Content. In 2012.

21. IMAGES WORKUP [Internet]. [cited 2019 Oct 22]. Available from: https://www.infochem.de/information/chemdraw-workup

22. May J, Lowe D, Sayle R. Sketchy Sketches: Hiding Chemistry in Plain Sight. In 2016.

23. IUPAC - International Union of Pure and Applied Chemistry [Internet]. IUPAC - International Union of Pure and Applied Chemistry. [cited 2019 Oct 22]. Available from: https://iupac.org/

24. OpenSMILES [Internet]. [cited 2019 Oct 22]. Available from: http://opensmiles.org/

25. Weininger D. SMILES, a chemical language and information system. 1. Introduction to methodology and encoding rules. J Chem Inf Comput Sci. 1988 Feb 1;28(1):31–6.

26. Daylight [Internet]. [cited 2019 Oct 22]. Available from: https://www.daylight.com/

27. OpenSMILES Specification: Table of Contents [Internet]. [cited 2019 Oct 22]. Available from: http://opensmiles.org/spec/open-smiles.html

28. Welcome to BlueObelisk.org [Internet]. blueobelisk.github.io. [cited 2019 Oct 22]. Available from: https://blueobelisk.github.io/

29. Daylight SMARTS - A Language for Describing Molecular Patterns [Internet]. [cited 2019 Oct 22]. Available from: https://www.daylight.com/dayhtml/doc/theory/theory.smarts.html

30. RDKit [Internet]. [cited 2019 Oct 22]. Available from: http://rdkit.org/

31. Chemistry Development Kit [Internet]. CDK - Chemistry Development Kit. [cited 2019 Oct 22]. Available from: https://cdk.github.io/

32. Open Babel [Internet]. [cited 2019 Oct 22]. Available from: http://openbabel.org/

33. O'Boyle NM. Towards a Universal SMILES representation - A standard method to generate canonical SMILES based on the InChI. J Cheminformatics. 2012 Sep 18;4:22.

34. Faulon J-L, Collins MJ, Carr RD. The signature molecular descriptor. 4. Canonizing molecules using extended valence sequences. J Chem Inf Comput Sci. 2004 Apr;44(2):427–36.

35. Schneider N, Sayle RA, Landrum GA. Get Your Atoms in Order—An Open-Source Implementation of a Novel and Robust Molecular Canonicalization Algorithm. J Chem Inf Model. 2015 Oct 26;55(10):2111–20.

36. CTfile Format [Internet]. [cited 2019 Oct 22]. Available from: https://www.3dsbiovia.com/products/collaborative-science/biovia-draw/ctfile-no-fee.html

37. Ketcher [Internet]. [cited 2019 Oct 22]. Available from: https://lifescience.opensource.epam.com/ketcher/index.html

38. OpenBabel FP2 hash function [Internet]. GitHub. [cited 2019 Oct 22]. Available from: https://github.com/openbabel/openbabel/blob/master/src/fingerprints/finger2.cpp#L259

39. openspecs-office. [MS-DOC]: Word (.doc) Binary File Format [Internet]. [cited 2019 Oct 22]. Available from: https://docs.microsoft.com/en-us/openspecs/office_file_formats/ms-doc/ccd7b486-7881-484c-a137-51170af7cc22

40. openspecs-office. [MS-DOCX]: Word Extensions to the Office Open XML (.docx) File Format [Internet]. [cited 2019 Oct 22]. Available from: https://docs.microsoft.com/en-us/openspecs/office_standards/ms-docx/b839fe1f-e1ca-4fa6-8c26-5954d0abbccd

41. [MS-CFB]: Compound File Binary File Format [Internet]. [cited 2019 Oct 22]. Available from: https://docs.microsoft.com/en-us/openspecs/windows_protocols/ms-cfb/53989ce4-7b05-4f8d-829b-d08d6148375b

42. Office Open XML File Formats [Internet]. [cited 2019 Oct 22]. Available from: https://www.ecma-international.org/publications/standards/Ecma-376.htm

43. Extensible Markup Language (XML) 1.0 (Fifth Edition) [Internet]. [cited 2019 Oct 22]. Available from: https://www.w3.org/TR/2008/REC-xml-20081126/

44. Media Types [Internet]. [cited 2019 Oct 22]. Available from: https://www.iana.org/assignments/media-types/media-types.xhtml

45. Sekine S. NYU: Description of the Japanese NE system used for MET-2. In 1998 [cited 2019 Oct 22]. Available from: http://citeseerx.ist.psu.edu/viewdoc/similar;jsessionid=8E9D6C6BFD2D7F873E5489EF07625221?doi=10.1.1.206.1406&type=ab

46. Lamurias A, Ferreira JD, Couto FM. Improving chemical entity recognition through h-index based semantic similarity. J Cheminformatics. 2015 Jan 19;7(1):S13.

47. Jessop DM, Adams SE, Willighagen EL, Hawizy L, Murray-Rust P. OSCAR4: a flexible architecture for chemical text-mining. J Cheminformatics. 2011 Oct 14;3(1):41.

48. Leaman R, Wei C-H, Lu Z. tmChem: a high performance approach for chemical named entity recognition and normalization. J Cheminformatics. 2015 Jan 19;7(1):S3.

49. Campos D, Matos S, Oliveira JL. A document processing pipeline for annotating chemical entities in scientific documents. J Cheminformatics. 2015 Jan 19;7(1):S7.

50. Munkhdalai T, Li M, Batsuren K, Park HA, Choi NH, Ryu KH. Incorporating domain knowledge in chemical and biomedical named entity recognition with word representations. J Cheminformatics. 2015 Jan 19;7(1):S9.

51. Lu Y, Ji D, Yao X, Wei X, Liang X. CHEMDNER system with mixed conditional random fields and multi-scale word clustering. J Cheminformatics. 2015 Jan 19;7(1):S4.

52. Xu S, An X, Zhu L, Zhang Y, Zhang H. A CRF-based system for recognizing chemical entity mentions (CEMs) in biomedical literature. J Cheminformatics. 2015 Jan 19;7(1):S11.

53. Tang B, Feng Y, Wang X, Wu Y, Zhang Y, Jiang M, et al. A comparison of conditional random fields and structured support vector machines for chemical entity recognition in biomedical literature. J Cheminformatics. 2015 Jan 19;7(1):S8.

54. Lowe DM, Sayle RA. LeadMine: a grammar and dictionary driven approach to entity recognition. J Cheminformatics. 2015 Jan 19;7(1):S5.

55. Swain MC, Cole JM. ChemDataExtractor: A Toolkit for Automated Extraction of Chemical Information from the Scientific Literature. J Chem Inf Model. 2016 Oct 24;56(10):1894–904.

56. BioCreative - Track 2- CHEMDNER Task: Chemical compound and drug name recognition task [Internet]. [cited 2019 Oct 22]. Available from: https://biocreative.bioinformatics.udel.edu/tasks/biocreative-iv/chemdner/

57. Hawizy L, Jessop DM, Adams N, Murray-Rust P. ChemicalTagger: A tool for semantic text-mining in chemistry. J Cheminformatics. 2011 May 16;3(1):17.

58. Peter Selinger: Potrace [Internet]. [cited 2019 Oct 22]. Available from: http://potrace.sourceforge.net/

59. Eiblmaier J, Kraut H, Isenko L, Saller H, Loew P. Challenges in Next Generation Scientific and Patent Information Mining. In 2011.

60. Eiblmaier J, Kraut H, Hausberg S, Loew P. Extraction of structural information from ChemDraw CDX files: easy, or an underestimated, difficult challenge? In 2013.

61. Ruby Programming Language [Internet]. [cited 2019 Oct 23]. Available from: https://www.ruby-lang.org/en/

62. RubyGems.org [Internet]. [cited 2019 Oct 23]. Available from: https://rubygems.org/

63. PerkinElmer Notebook [Internet]. [cited 2019 Oct 23]. Available from: http://www.perkinelmer.com/category/notebook

64. Nokogiri [Internet]. [cited 2019 Oct 23]. Available from: https://nokogiri.org/

65. aquasync. ruby-ole [Internet]. 2019 [cited 2019 Oct 23]. Available from: https://github.com/aquasync/ruby-ole

66. CDX Format Specification [Internet]. [cited 2019 Oct 23]. Available from: https://www.cambridgesoft.com/services/documentation/sdk/chemdraw/cdx/index.htm

67. Brecher J. Name=Struct: A Practical Approach to the Sorry State of Real-Life Chemical Nomenclature. J Chem Inf Comput Sci. 1999 Nov 22;39(6):943–50.

68. Simplified Wrapper and Interface Generator [Internet]. [cited 2019 Oct 23]. Available from: http://swig.org/

69. Nguyen A. Ruby GEM for RDKit [Internet]. 2019 [cited 2019 Oct 23]. Available from: https://github.com/CamAnNguyen/rdkit_chem

70. Wuts PGM, Greene TW. Greene's Protective Groups in Organic Synthesis. John Wiley & Sons; 2006. 1112 p.

71. Krallinger M, Leitner F, Rabal O, Vazquez M, Oyarzabal J, Valencia A. CHEMDNER: The drugs and chemical names extraction challenge. J Cheminformatics. 2015 Jan 19;7(1):S1.

72. PubChem PUG REST [Internet]. [cited 2019 Oct 23]. Available from: https://pubchemdocs.ncbi.nlm.nih.gov/pug-rest

73. Ruby on Rails [Internet]. Ruby on Rails. [cited 2019 Oct 23]. Available from: https://rubyonrails.org/

74. PostgreSQL [Internet]. [cited 2019 Oct 23]. Available from: https://www.postgresql.org/

75. JavaScript [Internet]. MDN Web Docs. [cited 2019 Oct 23]. Available from: https://developer.mozilla.org/en-US/docs/Web/JavaScript

76. React – A JavaScript library for building user interfaces [Internet]. [cited 2019 Oct 23]. Available from: https://reactjs.org/

77. CML - Chemical Markup Language [Internet]. [cited 2019 Oct 23]. Available from: https://www.xml-cml.org/

78. ChemDraw JS [Internet]. [cited 2019 Oct 23]. Available from: http://www.perkinelmer.com/product/chemdraw-direct-chemdrawdi

79. Kotov S, Tremouilhac P, Jung N, Bräse S. Chemotion-ELN part 2: adaption of an embedded Ketcher editor to advanced research applications. J Cheminformatics. 2018 Aug 13;10(1):38.

80. JSON [Internet]. [cited 2019 Oct 23]. Available from: https://www.json.org/

81. Sheldon RA. The E factor 25 years on: the rise of green chemistry and sustainability. Green Chem. 2017 Jan 3;19(1):18–43.

82. Wegener I. Complexity Theory: Exploring the Limits of Efficient Algorithms. Springer Science & Business Media; 2005. 307 p.

83. Klopmand G. Concepts and applications of molecular similarity, by Mark A. Johnson and Gerald M. Maggiora, eds., John Wiley & Sons, New York, 1990, 393 pp. Price: $65.00. J Comput Chem. 1992;13(4):539–40.

84. Tanimoto. An Elementary Mathematical Theory of Classification and Prediction. 1958.

85. Martin YC, Kofron JL, Traphagen LM. Do Structurally Similar Molecules Have Similar Biological Activity? J Med Chem. 2002 Sep 1;45(19):4350–8.

86. Tang CW, VanSlyke SA. Organic electroluminescent diodes. Appl Phys Lett. 1987 Sep 21;51(12):913–5.

87. Nuyken O, Jungermann S, Wiederhirn V, Bacher E, Meerholz K. Modern Trends in Organic Light-Emitting Devices (OLEDs). Monatshefte Für Chem Chem Mon. 2006 Jul 1;137(7):811–24.

88. Kim DY, Lai T-H, Lee JW, Manders JR, So F. Multi-spectral imaging with infrared sensitive organic light emitting diode. Sci Rep. 2014 Aug 5;4:5946.

89. Liu T, Zhao J, Xu W, Dou J, Zhao X, Deng W, et al. Flexible integrated diode-transistor logic (DTL) driving circuits based on printed carbon nanotube thin film transistors with low operation voltage. Nanoscale. 2018 Jan 3;10(2):614–22.

90. Friederich P, Gómez V, Sprau C, Meded V, Strunk T, Jenne M, et al. Rational In Silico Design of an Organic Semiconductor with Improved Electron Mobility. Adv Mater. 2017;29(43):1703505.

91. Python.org [Internet]. Python.org. [cited 2019 Oct 23]. Available from: https://www.python.org/

92. Flask [Internet]. Pallets. [cited 2019 Oct 23]. Available from: https://palletsprojects.com/p/flask/

93. Molconvert [Internet]. ChemAxon DOCS. [cited 2019 Oct 27]. Available from: https://docs.chemaxon.com/display/docs/Molconvert

94. Stewart JJP. MOPAC: A semiempirical molecular orbital program. J Comput Aided Mol Des. 1990 Mar 1;4(1):1–103.

95. Ahlrichs R, Bär M, Häser M, Horn H, Kölmel C. Electronic structure calculations on workstation computers: The program system turbomole. Chem Phys Lett. 1989 Oct 13;162(3):165–9.

96. Anastas PT, Warner JC. Green chemistry : theory and practice. Oxford University Press; 1998.

97. Trost BM. The atom economy--a search for synthetic efficiency. Science. 1991 Dec 6;254(5037):1471–7.

98. Dicks AP, Hent A. The E Factor and Process Mass Intensity. In: P. Dicks A, Hent A, editors. Green Chemistry Metrics: A Guide to Determining and Evaluating Process Greenness [Internet]. Cham: Springer International Publishing; 2015 [cited 2019 Oct 22]. p. 45–67. (SpringerBriefs in Molecular Science). Available from: https://doi.org/10.1007/978-3-319-10500-0_3

99. Sheldon RA. Organic synthesis - past, present and future. Chem Ind. 1992;

100. Roschangar F, A. Sheldon R, H. Senanayake C. Overcoming barriers to green chemistry in the pharmaceutical industry – the Green Aspiration Level™ concept. Green Chem. 2015;17(2):752–68.

101. Beilstein Journal of Organic Chemistry [Internet]. [cited 2019 Oct 27]. Available from: https://www.beilstein-journals.org/bjoc/

102. Aken KV, Strekowski L, Patiny L. EcoScale, a semi-quantitative tool to select an organic preparation based on economical and ecological parameters. Beilstein J Org Chem. 2006 Mar 3;2(1):3.

Chapter 9. Appendix

9.1 ChemScanner Usage and APIs

9.1.1. Usage

To scan/extract a single CDX file

```
require 'chem_scanner'

cdx = ChemScanner::Cdx.new
cdx.read('/path/to/cdx/file')
# Get array of scanned Canonical SMILES
cdx.molecules.map(&:get_cano_smiles)
# Get array of scanned Reactions in SMILES
cdx.reactions.map(&:reaction_smiles)
```

There are 5 classes correspond to 5 supported file formats:

- CDX - `ChemScanner::Cdx`
- CDXML - `ChemScanner::Cdxml`
- DOC - `ChemScanner::Doc`
- DOCX - `ChemScanner::Docx`
- PerkinELN - `ChemScanner::PerkinEln`

9.1.2. API

Molecule

- Access "scanned" molecules

```
# Molecules - array of scanned molecules
cdx.molecules
# Get array of scanned Canonical SMILES
cdx.molecules.map(&:get_cano_smiles)
```

```
# Get one molecule
molecule = cdx.molecules.first
# Number of scanned molecules
cdx.molecules.count
```

- `Molecule` class:

```
# Canonical SMILES
molecule.get_cano_smiles
# Molfile
molecule.get_mdl
# RDKIT RWMol
(https://www.rdkit.org/docs/cppapi/classRDKit_1_1RWMol.html)
molecule.rw_mol
# Molecule label (bold text near molecule)
molecule.label
# Molecule text (molecule description)
molecule.text
# Molecule details (additional information from Perkin Elmer
ELN)
molecule.details
```

Reaction

A `Reaction` object consists of 3 groups of molecules: reactants, reagents, and products. Each group is an array of molecules, which each element is an object of `Molecule` class. In addition, some abbreviations, that belong to the reaction, are represented by SMILES. Those could be accessed via `reagent_smiles`

```
reaction        = cdx.reactions.first
# Access extracted structure group
reactants       = reaction.reactants
```

```
reagents       = reaction.reagents
products       = reaction.products
reagent_smiles = reaction.reagent_smiles
```

Further manipulation of each group would be similar to the `Molecule` class.

Reaction properties

The `Reaction` class has `description`, `yield`, `time`, `temperature`, and `details` properties. All these properties are extracted from the ChemDraw scheme, except the details field is extracted from the additional information of `PerkinELN`.

Reaction step

Some multi-step reactions can also be recognized. If a reaction is a multi-step reaction, the "steps" could be accessed like follows

```
# Get first scanned reaction
reaction = cdx.reactions.first
# Access first step
step = reaction.steps.first
step.number # Should be 1
step.description
step.time
step.temperature
# List reagents SMILES
step.reagents
```

Each step has these following properties: `description`, `time`, `temperature`, and `reagents`.

Supported File Formats

CDX, CDXML, PerkinELN usage, and API are described above. Their outputs are simple `molecules` and `reactions`.

DOC and DOCX classes are a little bit different. DOC and DOCX files can contain more than one embedded ChemDraw scheme, in which each embedded scheme is one CDX scheme. ChemScanner attempted to extract all of them and put them into one `Hash` map called `cdx_map`.

```
require 'chem_scanner'

doc = ChemScanner::Doc.new
doc.read('/path/to/doc/file')
doc.cdx_map.each do |key, cdx|
  puts cdx.reactions.map(&:reaction_smiles)
end

# Access all molecules in all CDXs
doc.molecules.map(&:get_cano_smiles)
# Access all reactions in all CDXs
doc.reactions.map(&:get_cano_smiles)
```

DOCX is a bit different, `ChemScanner` can extract the CDX together with its preview image within the documents.

```
require 'chem_scanner'

docx = ChemScanner::Docx.new
docx.read('/path/to/docx/file')
docx.cdx_map.each do |key, cdx_info|
  # Get the CDX scheme
  cdx = cdx_info[:cdx]
  puts cdx.reactions.map(&:reaction_smiles)
```

```
  # Preview images, used for ChemScanner UI
  img_ext = cdx_info[:img_ext] # Could be '.png', '.emf'
  img_b64 = cdx_info[:img_b64] # Base64 encoded of image
end

# Access all molecules in all CDXs
docx.molecules.map(&:get_cano_smiles)
# Access all reactions in all CDXs
docx.reactions.map(&:get_cano_smiles)
```

9.2 Curriculum Vitae

Thanh Cam An, Nguyen

Born 28.11.1988 in Hue, Vietnam

E-mail: caman.nguyenthanh@gmail.com

GitHub: https://github.com/CamAnNguyen

Professional Experience

05/2016 - present	**Scientist** Institute of Toxicology and Genetics, Karlsruhe Institute of Technology, Germany
07/2014 - 01/2016	**Research Associate** Petabi, South Korea
09/2012 - 02/2014	**Software Engineer** Cargigi, Vietnam
11/2011 - 09/2012	**System Engineer** VNG Corporation, Vietnam
08/2011 – 11/2011	**Software Engineer** MBM International, Vietnam

Education

03/2014 - 03/2016	**M.S. in Aerospace Information Engineering** Konkuk University, South Korea
09/2006 - 06/2011	**B.S. in Computer Engineering** Ho Chi Minh City University of Technology (HCMUT), Vietnam

Publications

1. (under revision) Nguyen, A.; Huang, Y.-C.; Tremouilhac, P.; Jung, N.; Bräse, S., CHEMSCANNER: Extraction and re-use(ability) of chemical information from common scientific documents containing ChemDraw files. *Journal of cheminformatics*.

2. Tremouilhac, P.; Nguyen, A.; Huang, Y.-C.; Kotov, S.; Lütjohann, D. S.; Hübsch, F.; Jung, N.; Bräse, S., Chemotion ELN: an Open Source electronic lab notebook for chemists in academia. *Journal of cheminformatics* **2017,** *9* (1), 54.

3. Huang, Y.-C.; Nguyen, A.; Gräßle, S.; Vanderheiden, S.; Jung, N.; Bräse, S., Addition of dithi (ol) anylium tetrafluoroborates to α, β-unsaturated ketones. *Beilstein journal of organic chemistry* **2018,** *14* (1), 515-522.

9.3 Acknowledgments

I would like to thank all those who contributed to the success of this work.

Nobody has been more important to me in life than the members of my family. I would like to express my deepest gratitude to my family. I would like to thank my parents, Thanh Loc Nguyen and Thi Quyt Le, and my brother, Thanh Thien Nhan Nguyen, who encouraged me to pursue my goals and dreams through their trust and emotional support.

I am extremely grateful to my wife, Vu Hoang My Truong, and my son, Thanh Nhat Khang Nguyen, for your warm love, patience, and endless support.

I am especially indebted to my supervisor, Prof. Dr. Stefan Bräse, for the opportunity to work on the new exciting interdisciplinary research as well as the great motivation during the entire Ph.D. study.

Besides my advisor, I would like to thank Prof. Dr. Ralf H. Reussner for your acceptance to become my Korreferent.

I would like to extend my deepest gratitude to Dr. Nicole Jung, who always provides keen insight and valuable guidance. Without her guidance and persistent help, this dissertation would not have been possible.

I wish to thank Dr. Pierre Tremouilhac, Dr. Yu-Chieh Huang, Pei-Chi Huang, Chia-Lin Lin, Serhii Kotov, and Jan Potthoff. It is fantastic to work as a team with all of you.

I would like to thank Sylvia Vanderheiden-Schroen, Simone Gräßle, Julia Kuhn, Dr. Anke Deckers, Dr. Patrick Hodapp, Dr. Nicolai Wippert, Dr. Steven Susanto, Jerome Klein, Jérome Wagner, and Laura Holzhauer. Your precious support and assistance help us to improve the Chemotion-ELN project significantly.